Communist Entrepreneurs

Communist Entrepreneurs

Unknown Innovators in the Global Economy

John W Kiser III

FRANKLIN WATTS
New York Toronto 1989

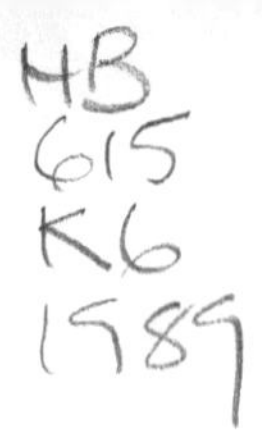

Photographs courtesy of: Oleg Homola: p. 72; ČSAV: p. 82; the author: pp. 121, 123, 128; The Walt Disney Company: p. 154.

Library of Congress Cataloging-in-Publication Data

Kiser, John W.
Communist entrepreneurs: unknown innovators in the global economy / John W. Kiser.
p. cm.
Includes bibliographical references.
ISBN 0-531-15110-7
1. Entrepreneurship—Europe, Eastern. 2. Technological innovations—Europe, Eastern. 3. Entrepreneurship—Soviet Union. 4. Technological innovations—Soviet Union. I. Title.
HB615.K6 1989
338.04′0947—dc20 89-36781
CIP

Printed in the United States of America
6 5 4 3 2 1

To Linda, David and Anne

Nothing universal can be rationally affirmed on any moral or on any political subject. Pure metaphysical abstraction does not belong to these matters. The lines of morality are not like the ideal lines of mathematics. They are broad and deep as well as long. They admit of exceptions. These exceptions and modifications are not made by the process of logic, but by the rules of prudence. Prudence is not only the first in rank of the virtues political and moral, but she is the director, the regulator, the standard of them all. Metaphysics cannot live without definition; but prudence is cautious how she defines.

EDMUND BURKE

Table of Contents

Acknowledgments

There are many people who have made this book come about. The initial impulse was provided by my old friend, Ned Chase, a New York editor who talked me into writing about my Eastern bloc exploits just at the time I was starting my company in earnest.

Numerous friends and officials in East Germany, Hungary, Czechoslovakia, and the Soviet Union deserve a collective thank you. All my interviews were conducted outside press channels, arranged mostly through existing business contacts who understood my dual status as writer and technology prospector.

Susan Eisenhower helped me gain access to Georgie Arbatov, whose name and blessing for the project got me quickly to the people I wanted to see in the USSR. Michael Rae of Argus International, with his impeccable Russian and interpretive skills, made my meetings with the machine tool magnate Vladimir Kabaidze truly informative.

A generous grant from the International Research and Exchanges Board (IREX) helped finance some of my travel expenses.

I owe a lot to my friend and literary mentor Willy Warner, who helped me grasp hold on a style and form of organization. My editor, Kent Oswald, gave me much needed line-by-line, chapter-by-chapter comments and critique, a service I'm told is becoming obsolete in the industry. He was a pleasure to work with, as was Julian Bach, my agent, who got me on the hook to produce a manuscript. The execution of the task into machine-readable form would not have been possible without my longtime secretary and cryptologist, Barbara Bradshaw.

Communist Entrepreneurs

Introduction

Every salesman, manager, engineer or member of an operating team is so used to thinking well of his own product, it is hard for any of them to recognize anything but the faults of his competitors.

HAROLD GENEEN

This book is about innovative individuals in Eastern Europe and the Soviet Union. I call them entrepreneurs, because they have traits commonly associated with entrepreneurship. They are achievement-oriented, and strong-willed, and they disregard rules. They are mavericks who have built productive and creative organizations in environments which reward people more for doing what they are told than for exercising initiative.

These entrepreneurs also represent "types" whose Western counterparts reside not only in industry but in government and universities. Hyman Rickover was a military-minded entrepreneur. He had a vision and fought tenaciously for its realization but was not motivated in his struggle by money. Lawrence Livermore Laboratory is the child of Edward Teller, a defense science entrepreneur motivated by ego and ideology.

Having a vision, selling it, organizing resources, and taking risks to realize one's goals are general characteristics of entrepreneurial personalities, regardless of motivation. In the United States, if such people find themselves locked in an unresponsive bureaucracy, be it IBM or Naval Research Laboratories, there is somewhere else for them to go to sell their ideas. Until recently, this has not been the case in Soviet-style economies.

The entrepreneurs whose stories I have presented fall into two broad categories: inventors and reformers. The "inventors" are those who are initially motivated by technical creativity, such as Otto Wichterle, the Czech chemist who first developed soft contact lenses, and Manfred von Ardenne, the East German physicist whose institute has supplied advanced electron beam melting furnaces to Sandia National Laboratories in New Mexico. The "reformers" are trying to change the system in some small way, usually by creating their own subsystem that operates by different rules. Elizabeth Birman was a reformer who wanted to create new mechanisms for financing technical innovation in Hungary. On behalf of better quality and service to his customers, Vladimir Kabaidze fought the Soviet system by ignoring its normal top-down dictates and writing his own rules.

In practice, the line between the two groups blurs, as the inventors may have to become reformers in order to realize their ambitions. Svyatoslav Fyodorov has built a commercial and technical empire in Moscow based on new financial incentives resulting from his drive to provide more efficiently to Soviet citizens the advanced ophthalmic surgical procedures he pioneered. Manfred von Ardenne used unorthodox methods to achieve his successes in electron physics, methods which became standard practice at his private institute in Dresden, East Germany.

What do these people represent? Why are they important? First, for me they represent inspiring stories of human achievement. Like that of salmon fighting upstream to lay their eggs, the stubborn drive of Soviet bloc entrepreneurs seems to be powered by mysterious genetic forces to fructify the world. There is little material incentive in the economic environment to reward the struggle. Successful entrepreneurs in the Soviet Union and Eastern Europe can live well compared to their fellow citizens, but a pervasive ethic of economic egalitarianism makes undue display of wealth bad form. Compared to their Western counterparts, they live like paupers. Not money, but variable mixtures of ego, idealism, and upbringing are the principal ingredients of motivation.

Second, these are success stories not in ballet or sport, but in spheres of achievement not normally associated with Communist countries: scientific innovation, technology, and management. Moreover, these entrepreneurs are from the era of stagnation, as the Brezhnev period is now called, active before *perestroika* unfolded in the Soviet Union and when Hungary alone was considered the daring darling of market socialism. As successful organization builders and fighters with proven ability to motivate others, they represent an overlooked part of the world stock of case histories in successful innovation.

Third, the eleven entrepreneurs in this book put a human face on systems whose economic and technical performance is normally described abstractly by dubious numbers and generalities. They represent the human capital that the systems of the Soviet bloc must utilize and organize much better in order to progress.

The three Soviets, Albert Kauls, Vladimir Kabaidze, and Svyatoslav Fyodorov, are proven leaders with well-articulated philosophies relevant to any audience. Otto Wichterle of Czechoslovakia and Manfred von Ardenne of East Germany are world class R&D directors who have created internationally recognized centers of excellence and whose technical innovations have been commercially applied worldwide.

The entrepreneurs in this book were selected most often on the basis of the areas where my own footpath had led me in Eastern Europe and the recommendations of friends in the countries in which I had concentrated my business activities. The Soviet Union, Czechoslovakia, Hungary, and East Germany have been the most active technology centers for Kiser Research. Therefore, it was in those countries that I most frequently encountered the types of people whom I have called entrepreneurs. These types exist in Poland, Bulgaria, and Romania too, but I have chosen to focus the scouting activities of our company in those countries where we believe there is the most opportunity and where we have established good working relations. There is no doubt that until recently the greatest scope for entrepreneurs has been in Poland and Hungary. Additional examples of entrepreneurship could be supplied more readily from those two countries than from the others. Today the Soviet Union offers growing opportunity for individual initiative but faces tremendous inertia and inbred cultural resistance, especially among the Slavic portions of a population accustomed to following orders. East Germany and Czechoslovakia have been skeptical observers of *perestroika*. Romania dances to its own drummer, while Bulgaria pursues a form of pseudo-*perestroika*. I have not sought, therefore, to paint a perfectly representative country-by-country portrait of either technological achievement or entrepreneurship behind the Iron Curtain.

Rather, these stories have a generic significance that is much greater than any statistical sampling would provide. From the successes of these unusual people, four broader themes are developed: (1) the challenge their innovations and lives pose to the stereotype of technological backwardness and inhumanity in Communist countries, (2) the business opportunities for alert Westerners who are looking for advanced technology, (3) the importance of these "rule breakers" for *perestroika*, and (4) the commonality of engineering problem-solving requirements and entrepreneurial behavior with political leadership.

These four themes are present more in the form of marbling in the cake than as principal constituent or distinct layers.

Before encountering them again in a dispersed state, I have some concentrated observations on each.

Challenge to Stereotypes

It has been convenient during the past forty years of hostility toward the Soviet Union to believe that nothing about their system is good. The phrase "technology gap" has become a national shorthand memo which fortifies America's belief in the inherent inferiority of the Soviet system, a belief which can lead the uninformed to a similar judgment about the people who live there.

It is a widely held notion, as far as science and technology are concerned, that communism causes brain death. Theft and collusion with unscrupulous Western companies are the only means left to the intellectually impoverished Soviet bloc for staying abreast, according to this worldview. Many creative people did indeed leave Russia after the Bolshevik Revolution. The combination of Hitlerism and communism had a devastating effect on the intellectual ranks of Eastern Europe. But as the story of the East German Manfred von Ardenne or that of the American physicist J. Robert Oppenheimer and many of his Los Alamos colleagues illustrates, not all scientists lean to the right.*

There are scientists and engineers of high standing the world over who have been and continue to be drawn to the ideals of economic and social equality. Visiting scientists and technical specialists from the Eastern bloc do periodically defect to the West, yet the vast majority return home, including world-class scientists such as Otto Wichterle of Czechoslovakia, who could be a wealthy man if he chose to live

---

*Oppenheimer was widely praised for his leading role in building America's first atom bomb. Yet during the McCarthy hysteria of the early 1950s, he was declared a security risk by the Atomic Energy Commission. His close association with Communists and fellow travelers, many of whom were brilliant physicists employed by him at Los Alamos during the war, constituted part of the case against him.

in the United States. Whatever the mixture of practical and philosophical reasons for returning willingly to barbed wire socialism, such scientists are part of an important Western heritage.

Belief that private accumulation of property is evil, but economic and social equality is good has a legacy stretching back through the French Revolution and numerous Christian thinkers to the Greeks. In Plato's *Republic*, the philosopher-kings were not allowed to own property or acquire wealth, lest their judgment be corrupted. The term *Reds* was first coined to describe the Jacobins of the French Revolution, not the Bolsheviks, who were much influenced by French radical socialism. The life of Jesus is not a story about the virtue of private gain. To pronounce that "Western" culture means pluralistic political democracy and free market capitalism to the exclusion of other ideals is to read history very selectively.

Yet even if all the technically oriented gray matter were not bred out under communist rule, it is obvious that the system has not made good use of its overabundant technical manpower. Lack of goods, poor quality, and inability to compete in Western markets bespeak not a lack of brainpower, but a system that has failed to provide incentives to produce high-quality goods and services. Not distinguishing between what the technical establishment in the Soviet systems knows how to do and what the system normally produces leads to a persistent tendency to underestimate them and produces attendant embarrassment for the United States.

Lacking the discipline of a commercial marketplace, product quality and performance tend to be driven by national security needs. Consequently, their technological achievements often are not visible to the casual observer, as they are not present in the consumer sector. The hundreds of millions of dollars wasted in constructing a now compromised American Embassy in Moscow was a result of grossly misjudging the level of Soviet sophistication in sensor technology. The hundreds of ceramic devices embedded in the embassy walls were virtually invisible to X-ray scanners. American specialists are still confounded by the combination of sophisticated software and remote sensing capability that enabled Soviets to monitor in real time the information being typed on the Embassy's IBM Selectric typewriters during 1986–87.

Technology includes raw technical invention, but value comes from commercial utilization and diffusion of the new technology into the economic life of a society. Not only are the incentives to use new technology weak in a society without competitive pressures, but poor quality is not punished in a system where buyers have no choices and producers are not allowed institutional death. Furthermore, the system cannot readily use new knowledge without the proper testing equipment, appropriate materials, components, scale-up facilities, and manufacturing infrastructure to support it. The Soviet and East European industrial chefs can often be the first to come up with a significant new industrial recipe, as they did with thin film diamond coating technology for hardening semiconductor chips, but it might require advanced utensils or specialized ovens or control systems to be cooked to commercial standards. The systems do not have the natural behavioral response to produce the requirements for high-quality, high-volume production, especially for nonmilitary uses.

Opportunities Offered by Soviet Bloc Brainpower

Opportunity can come from having information others do not. The information that led me to establish a business around the notion that the Soviet bloc represents undervalued and underutilized intellectual assets came as a result of an unsolicited proposal entitled *The Potential for Commercial Technology Transfer from the Soviet Union to the United States* that I presented in 1975 to the Department of State's Office of External Research. Congress had just passed the Jackson-Vanik Amendment, denying nondiscriminatory tariff treatment to the Soviet Union and other Eastern bloc countries until their human rights behavior improved, with emphasis on freer emigration for Jews. This was also the period in which détente, Henry Kissinger's unfortunate characterization of a world with less superpower tension, was rapidly unraveling. In this atmosphere, the argument was being made across a broad front that détente was a one-sided arrangement, favoring Soviet interests. Exhibit A for the antidétente position was Soviet acquisition of American technology. The receipt of dollars was not considered a fair exchange for the United States.

It was in this atmosphere that I raised the proposition that perhaps the United States should be taking greater advantage of Soviet technology. Is not the best defense a good offense? The initial study and a subsequent one which examined Eastern European sales of proprietary technology to the United States persuaded me that there was plenty of useful knowledge to be gained from the Soviet bloc, as well as business opportunity based on acquiring rights to advanced technology. The little-known case histories of corporate America's acquisition of patents and trade secrets from the Soviet Union and Eastern Europe gave strong reinforcement to my skepticism about the conventional wisdom and fueled interest in involving myself in East-West affairs.

The practice of hospital suturing was revolutionized in the 1970s by the introduction of surgical stapling guns, a technology acquired from the Soviet Union by a start-up company in Stamford, Connecticut. U. S. Surgical Corporation is now the world leader in surgical stapling and a $250-million-company that has sent shock waves throughout the industry over the past fifteen years. Kaiser Aluminum and Chemical and other aluminum companies throughout the world have purchased rights to a Soviet technology for casting aluminum that eliminates conventional molds by levitating the molten metal in pulsed magnetic fields, thereby improving quality and reducing processing costs. Advanced Soviet welding technology has been widely applied to American railroads and subways by the Holland Corporation of North Chicago. Through creative arrangements with Leningrad Polytechnic, Perkin Elmer, a Connecticut-based world leader in atomic absorption spectrometry used for measuring trace elements at parts per billion, acquired specialized sample preparation accessories that made their equipment more sensitive and more competitive. The United States soft contact lens industry was founded on Czech inventiveness.

Nor are American companies the only ones using the technological achievements of the Soviet bloc. (Table 1.) Japan has helped itself to a wide range of Soviet technologies for the iron and steel industry, in addition to shuttleless weaving technology from Czechoslovakia and plasma welding and liquid crystal technology from East Germany. Joint marketing companies have been formed in Italy and Germany to sell Soviet technology.

SELECTED EXAMPLES OF SOVIET BLOC LICENSES SOLD TO WESTERN COMPANIES 1965–1987

Buyer	*Technology*	*Source*
	Metallurgy	
Rolls-Royce (England)	shell molds for investment casting	Soviet Union
Creusot Loire (France)	electroslag remelting	Soviet Union
Ulvac (Japan)	" "	" "
SKF (Sweden)	" "	" "
Cabot Corp. (U.S.A.)	" "	" "
Nippon Steel (Japan)	evaporative stave cooling of blast furnaces	Soviet Union
Andco Engineering (U.S.A.)	" "	" "
Nippon Kokan (Japan)	dry coke quenching	Soviet Union

Reynolds Aluminum (U.S.A.)	electromagnetic casting of aluminum	Soviet Union
Kaiser Aluminum (U.S.A.)	" "	" "
Multiarc Vacuum (U.S.A.)	titanium nitride coating for cutting tools	Soviet Union
Tellus Maskin (Sweden)	electrohydraulic cleaning of castings	Soviet Union
Aschenbach Buschhutten (West Germany)	bimetallic rolling	Soviet Union
Iwatani (Japan)	plasma arc cutting	East Germany

Chemistry

Montedison (Italy)	process for making polycarbonates	Soviet Union
Polydex (Canada)	chromatography sorbents	Czechoslovakia
Ashai Chemical Industries (Japan)	process for making a building material silicalcite	Soviet Union
Hoechst (West Germany)	technology for increasing the production of alkyd resins	East Germany
Dow Chemical (U.S.A.)	process for chemical curing of unsaturated polyester resins	East Germany
Buna Engineering Ltd. (Canada)	silicone products	Czechoslovakia
Chromatronix (U.S.A.)	system for thin layer ion exchange chromatography	Hungary
Vekamef (Holland)	epichlorhydrin production	Czechoslovakia

	Food Processing	
Möet et Chandon (France)	automatic wine champagnization process	Soviet Union
Miles Laboratories (U.S.A.)	cheese fermentation enzyme	Poland
	Biomedicine	
U.S. Surgical Corp. (U.S.A.)	surgical staplers	Soviet Union
Bausch & Lomb (U.S.A.)	collagen eye shields	Soviet Union
	intraocular lens implants	Soviet Union
" "	soft contact lenses	Czechoslovakia
Compudrug USA	computer assisted drug and chemical design	Hungary
Tieman (West Germany)	cardiovascular agent Halidor	Hungary
Ferring (Sweden)	Adiuretin-SD for treating diabetes insipidus	Czechoslovakia
Automatische Mess u. Steuertechnik GmbH (West Germany)	Evolite lamp for treating chronic skin wounds and ulcers	Hungary
Diversified Technology Inc. (U.S.A.)	biodegradable orthopedic pin	Soviet Union
National Patent Development Corp. (U.S.A.)	instrument for customized eyeball measurement	Czechoslovakia
General Electric (U.S.A.)	compact flywheel energy storage system for use in X-ray machines	Hungary
U.S. Company	skin test for cholesterol	Soviet Union

Du Pont (U.S.A.)	cardiovascular drug	Soviet Union
	Miscellaneous	
Chisso Corp. (Japan)	liquid crystals	East Germany
Toshiba Ceramics (Japan)	method for growing monocrystals of corundum from the melt	Soviet Union
Maxwell Laboratories (U.S.A.)	magnetic impact bonding	Soviet Union
Texas Utilities Services, Inc. (U.S.A.)	*in situ* coal gasification	Soviet Union
Sumitomo (Japan)	pneumatic pipeline transport	Soviet Union
Mitsui Miiki Machinery (Japan)	OMTK long wall complex	Soviet Union
Air Products (U.S.A.)	separation columns	Soviet Union
Du Pont (U.S.A.)	textile machinery device	East Germany
Deering Company (U.S.A.)	injection spray printing of designs onto carpets	East Germany

It is Kiser Research's business to find advanced technology in the Soviet bloc countries and to arrange for its evaluation and acquisition by industrial clients or for its own new venture portfolio. Known as technology brokering, we marry two parties who need to work together technologically on terms as intimate as those of the marital chamber. Success in transferring detailed technical information and unwritten know-how, residing only in the heads of one's partners, requires good communication and diplomacy. As a technology broker, Kiser Research does not deal in products, but with the underlying recipes and intellectual property rights for making products or improving processes.

Broadly, technology is knowledge directed to solving real-world problems. Proprietary technology represents a special form of knowledge believed to provide a competitive edge. It can be protected by a patent or treated as a trade secret. Herbert Henry Dow, founder of Dow Chemical Corporation, implicitly defined proprietary technology when he asked his managers the feisty question, "If you can't do it better or cheaper, why bother?"

Science, in contrast to technology, represents a body of knowledge directed to understanding better what makes nature tick without regard to utility. Science is supposed to be predictive, so its knowledge is expressed as laws, such as Boyle's law of gases or Newton's laws of motion. Basic and applied science represent the intellectual capital account against which industry's engineers draw in order to develop innovative solutions to todays problems.

There is a wealth of open scientific and technical literature to be tapped for the price of a translation. The eighty-thousand-odd Author's Certificates (Soviet equivalent to U.S. patents) published annually in the USSR alone, together with the thousands of books and journals published each year, represent a large and substantial source of public knowledge which can be accessed. Currently, this resource is poorly mined, thanks to a general lack of awareness and the weak foreign language skills of the American technical community. Greater familiarity with this open knowledge can provide new sources of intellectual stimulation for solving technical problems and lead to more ventures and licensing of Soviet technology. The United States could reap many more of the kinds of benefits some of its alert corporate scientist have already gained.

In the 1960s, for example, Du Pont changed its production techniques for making a chemical intermediate, vinyl acetate, on the basis of a cheaper method described in the Soviet literature. AT&T researchers first got the idea to use lasers for annealing semiconductor chips from reading Soviet journals and hearing Soviet scientists present papers in Albany, New York, in 1977. Varian Associates' 1983 annual report credited Soviet research with helping to break a technical bottleneck in the development of ultrahigh-frequency gyrotrons (powerful millimeter radio wave generators used for communications, linear particle accelerators, and initiation of nuclear fusion reactions). Researchers at Los Alamos National Laboratory incorporated ideas from the Soviet literature to design more effective technology for neutral particle beam weapons based on radio frequency quadrupole. Beryllium is now added to American rocket propellants to increase thrust, thanks to Russian publications on the subject. Because it is a highly toxic substance, Americans had not considered its use. These are but a few examples of how past awareness of foreign work, and Soviet research specifically, has helped U.S. research efforts.

Eastern bloc scientists can be valuable resources precisely because they do operate in a different environment and are somewhat isolated from the West. The positive side of this isolation is demonstrated by original approaches derived from having a different set of constraints and preconceptions, or simply from reading different books. In the chemical engineering laboratories of Robert Cohen at MIT, a visiting Polish researcher, Andrzes Galeski, made an important contribution to the understanding of deformation processes occurring in nylon under physically stressed conditions. Thanks to some experiments he could undertake more effectively in Cambridge than in Lødz, where his laboratory lacked the computer-automated equipment and rapid data-reduction capabilities, he could prove his hypothesis. However, the stimulus to think differently from his American counterparts about this phenomenon came from his familiarity with East German writing on the subject. Galeski gained through access to hardware not readily in use in Poland; the U.S. polymer community gained a new insight into the behavior of a very important, widely used material.

The positive side of deprivation is the stimulus to innovation it can provide. Necessity, goes the saying, is the mother of invention. The shortage of chemicals and modern research instruments certainly hinders research but also fosters a tremendous amount of ingenuity in Eastern bloc scientists, leading to an emphasis on very simple but effective solutions. In contrast to the United States, where hands-on laboratory skills are dying out, the Eastern bloc has many excellent synthetic chemists—people who can make a chemical from scratch. These people are sought after by leading U.S. universities. Thus, much inventiveness is expended in the direction of finding solutions which permit the use of available material or equipment without sacrificing performance. In such cases, the solution may use a cheaper available material or simpler process. U.S. industry is seeking to make products that save labor, material costs, or investment in the face of a technical culture which emphasizes expensive "sophistication." Driven to find simple effective solutions in an environment of shortages, the "unsophisticated" ten-cent solution to a million-dollar problem is a tradition many in U.S. industry would like to practice more.

Opportunity most readily arises from talking face-to-face with other people. Only through firsthand contact does it become possible to find out what can be accomplished. Many of the people described in this book embody the opportunities that exist for those willing to have their preconceptions challenged. Much more than the now small trickle of technology would be benefiting the United States if there were greater recognition that Westerners also can learn from the Soviets and Eastern Europeans by looking at their strengths, not just their weaknesses.

The Soviet bloc system generally loses an R&D advantage for the same reason the U.S. Post Office would if it should produce a "first" by virtue of the curiosity of its employees.

The systems do not have the incentives and agility to exploit the commercial use of important breakthroughs, which also may be hard to recognize as important. This Soviet weakness, vis-à-vis the West, is analogous to the growing problem U.S. industry has with Japan, but with a twist. We frequently do something first, but the Japanese exploit it commercially. VCR technology was developed at RCA Laboratories, but the Japanese have conquered the market because RCA prematurely abandoned the research as not having any future. For lack of accountability, the Soviet systems are prone to pursue dead-end research for too long or to pursue research for a long period then fail to apply their results because of the lethargy of the system. Others reap the benefits, as the Japanese are doing rapidly with the thin film diamond technology pioneered in the USSR for depositing diamondlike crystals on substrates for semiconductors and electronic components to improve their performance in high-temperature environments.

The United States is more likely to do something for too short a period and quit prematurely because no payback has been realized. The U.S. research fuse is often too short; the Soviet, too long. Only a small handful of U.S. companies today can afford research whose payback stretches beyond ten years. Yet tough, fundamental problems require fifteen to twenty years of steady plodding. Soviet and Eastern bloc research is increasingly under pressure to pay for itself and show practical results. Commercialization of results is central to this requirement, but until Soviet bloc industry is motivated to be more aggressive in the use of new R&D results, Western business and entrepreneurs will continue to be a symbiotic catalyst for transferring Eastern bloc applied research to practical results.

Entrepreneurs and Reform—The Need for Rule Breakers

The processes of change occurring in the Soviet Union today, although seemingly new, have precedents in Eastern Europe stretching back thirty years. The reforms and intellectual ferment that have broken through the traditional surface calm of the Soviet society have much to do with the recognition by inherently conservative but intelligent managers of the system that its political credibility, economic future, and long-term security interests necessitate drastic changes. The nature of the highly visible internal debates indicates a predictable diversity of opinion on the subject of how much, how fast, and by whom the changes should be carried out. In the process, the premises of doctrinaire socialism are being reexamined in the same manner as were many of the premises of doctrinaire capitalism that led to its modification in the 1930s.

In its efforts to chart new ground domestically, the Soviet Union benefits from the "reform laboratories" the Eastern European countries potentially represent, and draws upon an intellectual legacy of dissent with the central planning mechanism as a regulator of economic life. In both word and deed, Yugoslavs, Poles, Czechs, Hungarians, and East Germans have all sought ways to break out of the intellectual straitjacket of central planning, or Soviet rule, or both. The history of postwar Eastern European reform movements shows clearly that reform agitation that acquires an overtly anti-Soviet tone is doomed to fail, unless protected by geography, as was that of Yugoslavia. The relative successes of East Germany and Hungary in de-emphasizing the role of central planning in favor of the producing entities of the economy have been accompanied by political stability. The East Germans have emphasized decentralizing decision making, restructuring industry, and giving more autonomy to its main producing units, the Kombinats. Hungary has steered its reforms in the direction of more reliance on market mechanisms.

The individuals described in the following pages worked with and through their establishments, not against them. The entrepreneurs presented here embody not only what the Soviet and East European systems need in greater quantity but are people who would benefit any system. The infinitesimal number of such independent-minded "rule breakers" who are willing to risk their careers bears out the axiom that changes in quantity lead to changes in quality. The quality of life in the Eastern bloc reflects the hand of systems that have chosen to suppress the role of individual initiative and spontaneity in the name of collective social interests, notions of social justice, and maintenance of political control. Now that the USSR has taken up the themes of its own former dissidents and those of its satellites from days past, the Eastern European experimenters are themselves less endangered species.

Commonality and a Socialist New Deal

Societies, like biological organisms, have certain common features—self-defense needs, resource requirements, transformation of resources into usable forms, distribution processes, control functions, and expressive needs. How these social functions are carried out varies as widely as their analogues in nature seeking to survive and prosper by adapting to their local environment.

Societies and organizations are also like products: they perform according to their design. There is a basic commonality to the problem-solving processes of both worlds. Both must find compromises between competing values. Both processes, organizing a society and designing a product, produce solutions which are specific to a particular environment or marketplace. Both technical and social solutions to a given set of problems can rarely be transferred to other countries without adaptation to local conditions.

A simple illustration of "local conditions" is provided by the story of ten-thousand Biolam microscopes the Boston school system was interested in buying from Russia back in the 1970s. Biolam microscopes are widely used in Russian schools, and they were being offered at a very good price. However, Russian microscopes are designed with removable eyepieces. This would not do for Boston. School officials knew that American children would quickly find some other use for the eyepieces than quiet observation of amoebae. Boston asked the Russian trade organization, Mashpriborintorg, to modify the design for the more obstreperous American market. It was impossible. This would be like asking General Electric to make ten-thousand units of a nonstandard toaster.

Local conditions help explain why similar problems will be solved in different ways. The basic challenges to socialism today and the obvious experimentation of Gorbachev resemble Franklin Roosevelt's predicament during the 1930s. Faith in unbridled laissez-faire capitalism had been shaken. Soul searching, agitation, and discontent were its products, leading to a new definition of capitalism and the role of the state which has been now enshrined in American history as the New Deal. To the opponents of the New Deal, Roosevelt was a traitor to his class and a Communist. The miseries of the Great Depression planted the seeds of disillusionment that led many to look to socialism as an alternative.

Indeed, some of the older socialist entrepreneurs in this book, such as Otto Wichterle and Manfred von Ardenne, came to their leftist leanings for the same reason many Americans did: capitalism was not working, and socialism seemed to offer a better vision of how the world should be. Just as disenchantment with capitalism appeared to serve the interests of the communist movement, today it might appear that the reverse process is occurring. The West is experiencing *Schadenfreude*, thanks to the concessions being made to the free market philosophy born from the failures of socialism. The mistake of the communist movement in the 1930s was to write off capitalism, because it was thought to be historically passé. Similarly, it would be a mistake for the United States to write off socialism because of a belief in the wickedness of communism. Gloating over the growing interest in and demand for freer markets and greater freedom of choice in the Soviet bloc as an indication of the superiority of "Western" ideals only serves to obscure for how long many of the premises of socialism have been accepted in the West under the heading of the welfare state. Both systems need elements of the other.

The late Joseph Campbell, a noted student of myth and religion, spoke of two types of moral systems derived from mythology. One is solar; the other, lunar. *Solar morality* refers to a moral outlook which sees the world in black and white: a clear, simple struggle between good and evil. *Lunar morality* derives from observing the moon in its duality of lightness and darkness combined in one sphere. Lunar morality sees good and evil as an intertwined continuum of gradations, each part of the other.

The entrepreneurs of the East are not solar, but lunar. They do not view capitalism as good; nor do they view socialism as evil. They seek a creative combination of both value systems, recognizing merit and demerit in both.

Chapter 1

The Hungarians Are Martians

A practical engineer is someone who perpetuates the mistakes of the past.

THEODORE VON KARMAN

Until Mikhail Gorbachev came along, Hungary was the socialist sweetheart of the Western press. The attention devoted to Hungary stemmed from its leadership in experimenting with Western-style reforms, a process that spread out over twenty years, starting in 1968: reduction of ministerial control over producers, greater price flexibility, and promotion of more private enterprise. Recent reforms have pushed market discipline further. In 1988, subsidies for failing producers were drastically reduced. The long recognized need for more mobility of capital has led to an incipient bond market. In January 1989, a public stock market was introduced. Western financial interests are already organizing Hungarian investment funds.

The Soviet Union has been an interested observer of Hungary for many years. The satellite countries provide laboratories for the Soviets to study the effects of different economic and organizational reforms. They have noted with interest the success of Hungarian agriculture, its relatively well supplied consumer sector, its production joint ventures with Japanese, Danish, and Austrian firms. Whereas the Russians are still trying to develop legislation allowing for joint ventures, the Hungarians already have it and are now using joint venture fever to construct imaginative arrangements with the Soviets on a variety of different fronts from biotechnology and lithium processing to bus making.

The greatest reform, however, was one of political attitude under the leadership of János Kádár. Installed by the Soviets after 1956, Kádár kept firm party control by adopting what Waterman and Peters call, in *In Pursuit of Excellence*, a "loose-tight" strategy. Reversing the Stalinist dictum "He who is not with us is against us" to the milder "He who is not against us is with us" has engendered an atmosphere of considerable openness in the debate over economic policy and a generally relaxed political climate. Although the Hungarians welcome *glasnost* and *perestroika* as supporting the kind of change they themselves are pushing, Hungary's place in the sun—never large in the best of times—is now much diminished. The Hungarians I will be talking about have been trailblazers in their own society and for the Soviets' as well. At least one Hungarian is breaking new ground for the United States, too.

The Martians

A theory about the Hungarians circulated at Los Alamos during the early 1950s when America's hydrogen bomb was being developed. In John McPhee's *Curve of Binding Energy* Ted Taylor, one of the young physicists working on the project, says the theory was that the Hungarians were really Martians in disguise. They had left their own planet and come to earth eons ago, but had the misfortune of landing in what is now

Hungary. The inhabitants there were so savage and primitive that it was necessary for the Martians to disguise their evolutionary advances and adapt to the native environment, lest they be hacked to pieces.

Although generally successful at camouflaging themselves, the Martians had three characteristics that were too powerful to be suppressed completely: their propensity to wander, as displayed by the Hungarian gypsy; their weird language, which is hardly related to anything; and their unearthly intelligence. The evidence of the superior Martian intelligence was plain to see: Edward Teller, Eugene Wigner, Leo Szilárd, and John von Neumann were all Hungarian emigrants. Wigner had designed the first plutonium production reactors. Szilárd had been among the first to suggest that fission be used to make a bomb. Teller, possessing a thick Martian accent, "coinvented" the hydrogen bomb with the Polish mathematician Stanislas Ulam. Von Neumann developed the principle of digital computing using binary logic, still the basis of all computers in use today. His pioneering book on artificial intelligence, *The Computer and the Brain*, was written for the Yale University Silliman Lectures in the mid-1950s.

Other Martians whose cerebral power has emanated from the Hungarian plain are Dennis Gábor, developer of the theory of holography, and Albert Szent-Györgyi, the first to study the role of ascorbic acid in the body and to synthesize vitamin C. Both men won the Nobel Prize for their work. Georg von Békésy pioneered an understanding of the mechanism of hearing, winning a Nobel Prize in 1961. Peter Goldmark developed long-playing records and the first commercial color TV system at CBS Laboratories, where he worked from 1936 to 1971. An early rocket pioneer, Hermann Oberth, came from the Transylvanian hinterland. He, along with the American Robert Goddard and the Russian Konstantin Tsiolkovsky, presaged the era of space flight with their controversial and ridiculed experiments. A virtual god in NASA's pantheon of heroes is Theodore Von Karman, the man who put the California Institute of Technology on the aeronautics map. The Von Karman Award today is the most prized form of recognition that can be obtained in the field of aerodynamics. All of these men, with the exception of Oberth, were born in Budapest and received advanced degrees at European universities, Goettingen in Germany being a favorite of the physicists; many had their careers blossom in the United States, thanks to Hitler's racial idiocy.

One of the frontiers of technology today on which the Martians are again present is the field of artificial intelligence. Popularly known as AI, it has been viewed as the province of academic oddballs who were trying to make machines that would think like humans. Advances in both technology and understanding of the human brain are making AI something practical people are looking at seriously. At a recent conference, an IBM executive pointed to the coming acceptability of AI by noting that for the first time, the "blue suits" were outnumbering the "propeller heads" at AI meetings.

Machines have long been accepted as superior to humans for the purpose of crunching numbers and processing large amounts of data. Machines can also keep track of information better than forgetful, easily distracted humanoids. Yet the domain of reasoning is one which many jealously guard as a uniquely human function.

Today's challenge is to go beyond relatively straightforward mathematical thinking to create software that can reason from an inventory of knowledge or combine mathematical with qualitative reasoning. Such artificial intelligence software is known as an "expert system," as it seeks to capture the knowledge and reasoning rules of people with defined expertise. An expert system can even learn to modify its own knowledge base, using new information and knowledge. An expert system acts, therefore, like an expert, perhaps even better, as many experts lose their ability to learn once they've been dubbed "experts." The frontier of expert systems lies in the development of statistical and mathematical tools for dealing with uncertainty, inadequate knowledge, fuzzy thinking, and even intuition, as these qualities are all part of the experts' decision making.

Ferenc Darvas

Researchers at Compudrug Ltd., a private cooperative firm in Budapest, have been in the forefront of expert systems software. Their specialty is expert systems for chemistry, especially drug design, metabolism, and toxicology prediction.

The driving force behind the innovative expert systems technology coming from Hungary is a forty-six-year-old polymath, who draws on the combined traditions of Von Neumann in computer science and Szent-Györgyi in chemistry. Ferenc Darvas's modesty would make him feel uncomfortable with the comparison, yet his ability to combine a broad command of chemistry, the discipline in which he was trained, with working knowledge of advanced mathematics, mathematical logic, and computer programming partly explains his small company's ability to penetrate the interdisciplinary thickets of biochemistry, pharmacology, mathematics, statistics, logic, and computer science successfully.

For drug design, computers can aid chemists in selecting molecular structures most likely to have activity—such as anti-infective action—that may also have fewer side effects. Expert systems can screen large numbers of compounds that are sitting "on the shelf" to determine whether they are likely candidates for having some desired "activity."

The U.S. Environmental Protection Agency (EPA) is now a leading candidate for using expert systems to predict toxicity in new chemical compounds or their *metabolites*, by-products of biological processes which break down one substance into another. The application of artificial intelligence to these fields of use is one in which the Hungarians are a formidable intellectual force, now a commercial one. "I've always thought the Eastern Europeans were ahead in this field," exclaimed Dr. Gilman Veith of EPA when he heard that a Hungarian company already had a software system for predicting metabolites, something that he was about to spend hundreds of thousands to develop from scratch.

Compudrug, Ltd. is largely a family operation. Ferenc Darvas is one of nine siblings, five of whom work with him in the business. Founded in 1981, Compudrug was one of many companies established as a result of changes in the law that year that allowed a new form of private cooperative which could issue stock to its members, who could number up to one hundred. Cooperative members are required to invest money or in-kind contributions to the enterprise, for which they are issued stock of equivalent value. Members own the company and are responsible for its management. If the business does badly and cannot meet its payroll, members are expected to forego their salaries before their regular employees. The bright side, of course, is that the shareholders are the first to participate in profits and have no limit on the amount of profit they can distribute to themselves.

In Hungary today, there are over two thousand private cooperatives of all kinds in the software business, averaging thirty to forty employees. They fail at the rate of five to ten a month. Many were formed by entrepreneurial programmers who left their jobs and set up businesses in order to sell their services back to their former employers at higher rates. This unexpected development in the wake of the 1981 reforms was not uniformly cheered by government planners. The original intention was not to encourage employees to use the marketplace to drive up their rates, but to create new businesses. The validity of the criticism hinges on whether these "entrepreneurs" who sell their services to their former employers function more productively than when working as underpaid and undermotivated employees of state companies. If Darvas is any reflection of the multitude of new software entrepreneurs, the answer is a clear yes!

Ferenc Darvas was born to a middle-class family. His grandfather was a pharmacologist, his father a mechanical engineer. The sixty-six-year-old mother of the Darvas clan is studying religion at the Catholic seminary in Budapest. She is taking what amounts to an adult education course in theology so as to be able to provide some Christian learning and values to her twenty-five grandchildren.

Darvas studied at the famous Rákóczy High School named after an aristocratic family long associated with Hungary's struggle for national independence against the Turks and Hapsburgs. Even then, the school still had Catholic priests as teachers, but it had lost much of its earlier rigor. One chemistry teacher was so bad that Darvas used to stand up in class and correct him: not the way to get A's. In addition to being brash, Darvas was the only student in his class not a member of the Communist youth organization, KISZ. Lacking working-class origins important for advancement in the post-Stalinist 1950s, Darvas was not accepted by the university the first time around. Instead, he got a job as a laboratory assistant in the organic chemistry department of the Budapest Technical University for a year.

The year spent cleaning test tubes provided Darvas his working-class credentials, allowing him to be admitted to the university. Yet, even as a student, Darvas kept working as a laboratory technician and earned a reputation as being skilled at experimentation. Upon graduating in 1966, he got a job as a bench chemist at the EGYT Pharmaceutical Works. After several years, Darvas was promoted to the position of laboratory director and, as a result, became interested in a curious "Monday morning" anomaly. His group was responsible for the production of an antibiotic known chemically as chloramphenicol. In making a study of the yield of certain intermediate products needed for the synthesis of the antibiotic, Darvas noticed that the yields were higher on Mondays than other days of the week. In all parts of the world, Mondays are typically "low-yield" days, caused by the soporific effect of weekend slowdown on the productive juices. This made

Darvas wonder about the cause of such an unexpected finding. He had the idea that some "rusting" occurred over the weekends, when the production equipment was not in use. Thanks to the sensitivity of the production process, minuscule amounts of iron oxide on the equipment was all that was needed to act as a catalyst which speeded up the reaction process. Unfortunately, too much iron oxide would also cause an explosion.

This discovery of "rust" on his equipment as a potential multiplier of production yields sent Darvas to the library to try to figure out how to optimize the rust-induced catalytic reaction. The answer was found in some earlier mathematics developed by two Englishmen. They had worked out a method of numerical optimization to find a maximum of a mathematical function when the universe of the function is unknown. In other words, what is the optimum number of people who should be wielding a hoe to get maximum productivity from a plot of land, if you do not know the size of the plot? This problem is solved by the sequential simplex method.

Darvas took this mathematical finding and thought about it during his idle moments in the army reserve that summer, emerging with some novel ideas for applying this technique to chemistry. These ideas were first published in the prestigious *Journal of Medicinal Chemistry* in 1970. Darvas was aware of chemists' discomfort with computers, so he simplified the method, making it usable by anyone possessing pencil and paper.

Coincident with his thoughts in sequential simplex was Darvas's growing disenchantment with organic chemistry. After earning a degree in patent law within six months by going to night school, Darvas decided that drug design was the most interesting field of medicinal chemistry. It combined practical utility and intellectual challenge, as computer-aided drug design requires a deep theoretical approach.

Armed with his new knowledge of optimization techniques, Darvas with four other chemists tackled the problem of increasing the yield of chloramphenicol, the synthesis of which was notoriously difficult, and impossible to transfer to large-scale production without serious impairment of quality control. Using sequential simplex to figure out the optimum production conditions, in conjunction with the discovery of a new catalytic system, Darvas and his coworkers obtained for their employer, EGYT Pharmaceutical Works, a patent for a high-yield production process that also allowed for adequate quality control. This technology was used to fulfill large Japanese orders, and at twenty dollars per pound, the initial ten-ton order produced a nice cash flow for EGYT.

Under Hungarian law, developers of new technology are entitled to a small percentage of the income stream. Even at 0.5 percent royalty, the Japanese orders were creating the potential for Darvas and his coinventors to earn more money than the managing director of EGYT.

This awkward situation provoked the director to ask the five patent holders to sign a waiver of their right to further payments. Three signed, and two refused. Darvas was one of the refuseniks. His act of defiance led to a barrage of daily harassment by the director, culminating in the accusation that he had even bribed the patent office to get his Author's Certificate, the Hungarian equivalent of a U.S. patent.

Fed up, Darvas decided to get a new job at the Computer Center of the Ministry for Heavy Industry, known as Nimiguszi. This change was accomplished without difficulty, partly because the bad reference was not taken seriously and partly because politics had not gained much of a role in the relatively new field of computer science.

Darvas considered Nimiguszi to be the most exciting place to work at the time for anyone interested in computers. It was here that he first learned about Prolog, the powerful language developed by a Scottish team for writing advanced software programs. Two of Nimiguszi's associates had spent some time in Glasgow and had just come back enthusiastic about Prolog and with the code for the Prolog compiler in their hands. The compiler acts as the translator between the program and the machine, converting human instructions into machine instructions.

While at Nimiguszi, Darvas headed up a group using Prolog to write drug design software. Such software helps drug companies identify, at an early stage, compounds that may be harmful and assists in selecting compounds that show promise for achieving the desired therapeutic effect. One of his first customers was none other than EGYT Pharmaceutical Works. The final irony came when the managing director of EGYT was himself fired for not putting out the proper number of flags around the factory on August 20, Hungary's national Constitution Day.

Why Compudrug?

Compudrug Ltd. is a rapidly growing, $1.5 million business, most of which is still derived from sales in Hungary. Through Compudrug USA, a marketing subsidiary, unique expert systems are now reaching the U.S. market to aid organic chemists, metabolic specialists, and toxicologists in industry, government, and universities. Knowledge that a nonspecialist in metabolic processes might need weeks to acquire by himself or that would simply be ignored because it was too difficult to obtain is now available within seconds. For the specialist, metabolic expert systems provide a check on his intuitive knowledge and a tool for controlling the thoroughness of his own work. EPA, Upjohn, Marion Laboratories, Lederle, Glaxo, and Bayer are some of the beneficiaries of Compudrug's knowledge.

Why have these breakthroughs in the commercial development of sophisticated state-of-the-art software occurred in Hungary? Why Darvas? There is no easy answer, but there are two general features of Hungary that are relevant.

One is tradition. The popular image of Hungary as the land

of gypsies, salami, and the Gabor sisters belies the lesser known reality of Hungary as a country with strong traditions in mathematics, computer science, and chemistry, all of which play an important role in Compudrug's expert systems. Hungary has produced a disproportionate number of world-rank mathematicians, such as Pal Erdos and Laszlo Lovasz, famous, respectively, for their work in set theory and combination theory. Mathematics continues to occupy a strong position in the curriculum of Hungarian schools. A solid grounding in mathematics allows a more rigorous approach to problem solving of all kinds. Without mathematics, the related subjects of mathematical logic, advanced statistics, and quantum chemistry are inaccessible. In a word, Hungarians get good training in fundamental science and its indispensable handmaiden, mathematics.

The second consideration is that political influence in the computer industry, especially in software, has been either absent or benign. Unlike heavy industry and traditional manufacturing, where Party membership has been an important condition for advancement, computer science is a new field, in which Party influence on professional decisions is minimal. For a long time, computers did not fall into the category of "the commanding heights" requiring Party domination, as did steel production and electrical power and other areas of obvious industrial significance to the military strength of a nation. The implications of computers were not fully understood, nor did the technology fit into Marxist paradigms of worker control over means of production. In the early days, computers were viewed skeptically, as Americans might regard ESP research.

Consequently, when Darvas left the pharmaceutical industry for the greener pastures at Nimiguszi, the Computer Research Center of the Ministry of Heavy Industry, his lack of Party membership and poor recommendation did not count for much. Even though Nimiguszi belonged to the Ministry of Heavy Industry, a traditional Party-dominated ministry, the institute itself was free of old-line ideological thinking. Quite the opposite. Its director was a freethinking, innovative person who was only interested in having creative people at his institute. It was here that Darvas learned about Prolog, well-adapted for writing complex programs based on deductive logic. One of the great advantages of Prolog over its older American competitor with the unfortunate acryonym LISP, meaning list processing, developed in the late 1950s by Stanford University computer scientists, is the relative ease with which a classical syllogism such as "All men are mortal: Socrates is a man: Therefore, Socrates is mortal" can be expressed.

From the days when Christian scholastics used Aristotelian logic to argue God's existence rationally through Sherlock Holmes of a later era, deductive thinking has been a powerful tool for arriving at reasoned conclusions from premises. It is something we all do every day, some of us better than others. By easily accommodating the expression of mathematical logic functions, Prolog is a language particularly well suited for reasoning, including the non-Aristotelian variety: fuzzy, inferential, uncertain.

Programs using this kind of logic can be written with relative ease in Prolog and require few lines of *source code*, the hieroglyphics upon which the program instructions are based. Where LISP requires one hundred lines of source code to express a logical concept, Prolog may require only ten. It is also a language adaptable for writing programs that require treating time as a variable: Three people are leaving for a rendezvous at the base of Mt. Fuji from three different parts of the globe. At what speed must they each travel to arrive simultaneously? Many problems of biochemistry and other areas require taking account of time. T Prolog, developed by Compudrug, offers a particularly good solution to the temporal dimension of events that must be accounted for to produce soundly reasoned conclusions.

Another part of the answer to the Compudrug achievement also lies in the European-American scientific rivalry. The not-invented-here syndrome, combined with strong vested interests in the LISP language developed by the American computer science establishment, resulted in Prolog's not being taken seriously. It was foreign. At first, it was ignored and then dismissed. Darvas is part of the European scientific network. Other colleagues in Budapest were the first to develop a commercial Prolog compiler, known as *M Prolog* (*M* for "Magyar"), which became the basis of a successful Canadian business.

Thus, early exposure to Prolog at Nimiguszi, coupled with ability to recognize its potential for chemical problem solving and drug design applications, has given Darvas and the Compudrug team more experience with this powerful AI language than any other group in the world. And, it is Darvas's ability to grasp Prolog's significance; teach himself computer science and programming; apply these tools to the subject in which he was trained, chemistry; and then assemble a first-class interdisciplinary team that accounts for much of Compudrug's technical success. For this reason, Darvas has been a featured speaker at IBM, Mt. Sinai Hospital, UCLA, and many other centers of interest in AI and software for biochemistry. The difficulty of the subject is evidenced by the relative scarcity of available scientific software for the chemical industry. Much of the software in this category listed in the *Encyclopedia of Scientific Software* has been written by Compudrug.

Most important to Compudrug's commercial lead over the U.S. artificial intelligence technology was the early focus on producing user-friendly solutions to problems on standard IBM-PC equipment, clones of which are readily available in Hungary. Early American efforts not only were too absorbed in the technology itself but delivered solutions in the form of do-it-yourself applications kits or "shells" that required expensive, specialized computers for their operation. The

American AI pioneers were basically technologists at heart and forgot that the market was not interested in technology but in results that were cost-effective and easy to produce.

Incentives for Innovation: Primacy of the Technical

In Hungary, as in other Eastern bloc countries, researchers stand to benefit materially from the successful commercialization of their work. By contrast, the typical corporate researcher in the United States is paid a comparatively decent salary but is expected to produce. As in Eastern Europe, his work belongs to his employer. He has no stake in his work other than ego, a powerful one to be sure. It is rare in America that researchers whose discoveries lead to extremely successful products get anything other than a token bonus. One reason for this is the same one that led to Darvas's resignation from EGYT. Most American CEOs cannot accept the notion of subordinates' making more money than they do. Ironically, American management may take a more collectivist view of the requirements of commercial success than many Eastern bloc organizations. Bringing a new product to market is typically regarded as a "team effort," American for "collectivism." Under this corporate culture, the brilliant researcher's breakthrough is only one link in a chain that includes contributions by engineering, manufacturing, marketing, and sales. Groupthink, brute-force research, combined with the primacy of the marketing over the technical mentality, adds up to a lesser degree of recognition for the importance of the inventive individual's contribution in corporate USA than in many Communist societies where marketing is still an alien concept.

In Hungary and other countries within the Soviet bloc, it is the engineering and technical culture that dominate. Pure technical accomplishment is valued for its own sake, also a German influence. This attitude is greatly facilitated by the lack of any need to market, sell, or otherwise struggle to woo a buyer confronted with competitive alternatives. Whereas in the United States during the 1960s "marketing" got a bad name as synonymous with psychological manipulation, it is precisely in the area of marketing that Hungary and other communist countries recognize one of their greatest weaknesses. They need to figure out how to produce what people really want, rather than simply meet production quotas. When a young Soviet visited Harvard Business School in the early 1970s, he shocked one of my left-leaning friends by saying that his country's greatest problem was lack of marketing know-how, synonymous in my friend's mind with flimflam.

The propulsive forces for innovation in Hungary come not from the marketplace, where the failure to be competitive has not been life-threatening in the past. The main drivers of innovation arise from the technical community, the people who believe their mousetraps really are better. But lack of strong competitive motivation by industry to use innovative research results in a large amount of R&D remaining "on the shelf" instead of getting in the factory. Or the physical infrastructure to produce the type of machine needed to make the new electronic circuit simply does not exist. And there is no incentive to fill small, specialized niche markets. Only recently in Hungary can the frustrated researcher have an alternative. He can leave his place of work and strike out on his own, as Darvas did. In doing so, he may receive help from Elizabeth Birman.

Innovation Fund

The development of a state-financed venture-capital industry in Hungary can be directly traced to an unassuming, fifty-year-old Hungarian Liza Minnelli look-alike. Elizabeth Birman comes from fighting stock. Her grandfather was Eugene

Landler, a lawyer, who made a name for himself in Hungary during World War I by defending striking railroad workers. He became an important figure in the short-lived Communist Republic of Béla Kun established in March 1919. Landler was appointed successively minister of welfare and then of defense in the brief 133-day life of the Republic.

After the fall of Béla Kun, Landler's remaining years were spent in Vienna directing underground worker movements in Hungary. After his death in 1924, his wife and daughter moved to Moscow. Landler's body was sent from the sanatorium in Cannes where he was being treated for heart disease when he died. His is one of five non-Soviet bodies entombed in the Kremlin wall, joining Rosa Luxemburg and John Reed.

Birman was born in Kramatorsk in the Ukraine, where her father was a Party secretary in the large Kramatorsky Machine Building Plant. Known by his family as a 200 percenter for his loyalty and dedication to the Party, Birman's father was, nevertheless, purged as an enemy of the people in 1937. He died in prison before ever seeing his daughter.

As we ate our cold cherry soup at the Gellert Hotel, Birman asked whether I had read *Darkness at Noon* by Arthur Koestler. "It is the most excellent description of the psychology of the period. It is really fantastic that a man like Koestler who had never directly experienced terror could capture so perfectly what happens to people."

Despite her father's liquidation, or possibly because of it, Birman considered her childhood in the Soviet Union indistinguishable from that of any other Soviet child. There was no sense of nationality in those days. Stalin was a god to her, as he was to all the other Soviet children. Ten years later, in 1946, Elizabeth and her mother, brother, and grandmother moved back to Budapest. Fluent in Russian, Elizabeth attended the Russian language school in Hungary. All subjects were taught in Russian except Hungarian literature, geography, and history. Birman is still grateful to the Russian school for its rigorous training in basic mathematics, a training which enabled even the dullest student to answer questions which stumped those from the regular Hungarian schools.

The sight of postwar Budapest lying in ruins made the idealistic young Birman want to help in its reconstruction. Her mother wanted her daughter to be an economist, like her, but Elizabeth's independent streak prevailed, partly because whenever she asked her mother what economists did, the explanations never seemed to make any sense. After the discovery that she had absolutely no talent for construction engineering, Birman got her first job, ironically, as an economist in the Ministry of Construction.

Today, Birman describes herself as a pragmatist. It wasn't always so. Reared in a strongly orthodox Communist household, she never questioned, never considered other alternatives to organizing society. Birman knew something about the reality of Communist life. At the age of fourteen, she began to suspect that her father had not died in the war after all. Why weren't there any pictures, memorabilia, or other evidence of his wartime career? One day Birman found her way into a chest in which her mother kept her old correspondence. Curious, she uncovered news clippings and drafts of letters her mother had meticulously composed before sending them to her incarcerated husband.

Birman's mother never became disillusioned with the Communist cause, despite her loss. She rationalized that her husband was a casualty of the war. Just as innocent soldiers can be hit by their own artillery fire, so, too, are innocent people caught up in society's internal struggles. The harsh realities of the terror served to inoculate Elizabeth against the disillusionment that led some believers in the Communist paradise to abandon the course when confronted with its brutal side.

Yet 1956 turned out to be a shock for Birman anyway. It had never occurred to her that so many people in her own country were opposed to Stalinism. She had accepted it, warts and all, as a way of life and naturally assumed others had too. The

Hungarian uprising made Birman realize that there might be different paths to achieve a better Communist society. Nineteen fifty-six marked the beginning of her drift toward pragmatism and shedding of dogmatism.

In 1971, Birman left the Institute for Economy and Organization and was employed in the foreign trade department of the National Bank of Hungary. The principal function of this department was to provide credit to export-oriented businesses. One of their accounts was a local company called Vepex (meaning vegetable protein extraction). With Alfa Laval of Sweden, Vepex was developing refinements to a technology created in Hungary for the production of animal protein from cheap grasses that could be grown in sub-Saharan and similar inhospitable climates. The pacifist Swedes had dubbed the Vepex process the "green atomic bomb." When Janos Fekete, the deputy president of the bank, returned from a trip to Sweden, he passed on this catchy phrase to the Hungarian media. Impressed by the Swedish reaction to the potential importance of the Vepex process to solving world hunger problems, Fekete also told Birman to look further into the business and the chemical process.

Not long after his Swedish visit, Fekete was interviewed on TV. During the program he discussed the "green atomic bomb," saying that the National Bank supported innovation and supported Vepex specifically. The news that the bank supported innovation by its financing of Vepex produced a torrent of letters from other would-be innovators needing financial support.

From this was born in Birman's mind the idea that the bank should support technical innovation. But how? For several unsuccessful years, Birman and her boss, Soltesz, tried pushing different approaches that didn't succeed, such as offering manufacturers credit with which to finance the commercialization of their new inventions. Manufacturers never became very excited by this proposal, as it simply entailed paying for the privilege of adding risk to their business.

In 1979, a provocative article entitled "Are We So Rich?" appeared in a widely read publication, *Life and Literature*, known popularly as "And." Written by the poet Andras Mezei, the piece provoked tremendous discussion and controversy. The international success of Rubik's cube had caused Mezei to ask why this need be an isolated case. For months, letters and articles were written on the subject of innovation: what is it? how to foster it?

After several months of debate in the pages of "And," all the participants were asked to meet at the Hungaria, a well-known restaurant and meeting place of literati. Birman broke off her vacation at Lake Balaton, just southwest of Budapest, to take part in the sequel to the "And"-sponsored debate. There she met an old schoolmate, Borisz Szántó, who was working in the Secretariat of a parliamentary Committee for Science Policy. This committee had ideas about the promotion of innovation and was thinking in parallel with the discussion which took place at the Hungaria Restaurant. Szanto proposed that Birman join an ad hoc committee of the State Office for Technical Development, known as OMFB. OMFB's role is to develop broad guidelines and directions for encouraging and promoting industrial innovation. The ad hoc committee of OMFB was to prepare recommendations to the government's Science Policy Committee.

Throughout her time on the committee, Birman had been studying Western examples of state-supported innovation organizations, especially in Austria, France, and the United Kingdom. Her concept of creating a joint venture with a Western firm to invest in Hungarian technologies received a cool reception. Her other idea of creating a pool of risk capital to invest in new technology went not only against the grain of the bank but against the risk-avoidance psychology permeating the society. A quote in the British magazine *Puzzler* captured for Birman the bank's conservatism. "A bank is an institution which loans you an umbrella in sunny weather and takes it away when it rains."

Agonizingly long discussions on the ad hoc committee about innovation, what it is, and how to encourage it, paralleling the Hungaria Restaurant debate, produced three approaches: (1) create a committee of experts to review Hungarian innovations and make recommendations; (2) provide more money to two existing domestic organizations, Novex and Licencia, whose job was to promote the licensing of technology, both to domestic users and to foreign companies; and (3) create a pool of risk capital to be invested in start-up situations.

The committee approach had the obvious disadvantages of all such arrangements. Neither Novex nor Licencia liked the idea of being responsible for investing government money. Novex was directed by a risk-adverse woman whose reputation as a decisive manager was established at a conference when she fired her deputy for not keeping track of her handbag while she used the powder room. Birman's concept of a pool of risk capital for investment in new technology ventures had the merits of originality and clear responsibility for results.

Under pressure to produce a report for the government's Committee on Science Policy, the OMFB representative on the committee drafted recommendations to the board of OMFB. Birman saw the draft in advance and was bitterly disappointed that there was no mention of her ideas at all, not even a sentence. The next day, she was surprised to hear that her recommendations had been accepted.

According to the story, Lenard Pal, the OMFB president, hit the roof when he was served up a bowl of bland mush from the ad hoc committee: a little bit of this, a little of that. Pal wanted to make strong recommendations to the government to do something. The OMFB representative on the ad hoc committee who concocted the cream of wheat offered that there was an idea by Birman to establish a venture fund at the national bank. Pal liked the idea and sent it over to the bank for approval.

In trying to figure out how to finance the fund, Birman knew that OMFB itself and many of the ministries gave direct grants to firms to support technical innovation. The money for these grants came from mandated accounts funded from the state budget to provide ministries with resources to support technological innovation at the manufacturing level. Manufacturing companies had to create similar funds from their revenues. In reviewing these accounts, she discovered that there were substantial unused balances. The venture-capital pool would be created from these unused balances.

Pulai, the first deputy president of the bank, liked Birman's proposal and called his lieutenants to a meeting to discuss it and enlist their support. The Innovation Fund, as it was to be called, would be funded one-third from the national bank, one-third from OMFB, and one-third from the ministries. Total capitalization was to be 600 million forints, or about $12 million.

With the National Bank's support nailed down and an Innovation Fund lodged under its roof, the OMFB presented its recommendation to the Committee on Science Policy chaired by Deputy Prime Minister George Aczel. In the milling around before the beginning of the session, one of his ministers asked Birman how she came up with the figure of 600 million forints for the fund. Birman recalled the Russian way of trying to determine the right amount of vodka to drink, which they measure by the gram: 100 grams is too little, 200 grams is too much, but twice 150 grams is just right. In such ways, budgets are determined the world over.

Despite opposition by some ministers who thought the Innovation Fund was a crazy idea, the concept had the support of other key people, including the deputy prime minister, Aczel. Initially, the Innovation Fund bore little resemblance to a capitalist venture fund, but compared to most Western banks in the 1970s the fund was an original concept and an innovation for Hungary.

It obtained 150 million forints ($3 million) from OMFB, but

the capital contribution of the ministries was not automatic. Birman had to go to them to persuade, cajole, and fight for her stake. The standard ministry line was to ask how she presumed to know how to direct money better than the ministries, especially for projects related to those industries. Birman's parry to their questions was always the same. An organization which spends money on projects in order to make money will spend it differently than one which receives the same money as a gift. Also, Birman argued, she supported individual innovation, not companies, unlike the ministries. In the end, Birman persuaded the different ministries to ante up a substantial portion of the required ministerial share, in addition to the National Bank of Hungary and OMFB.

During the first five years, the fund acted autonomously, though its fixed costs were fully absorbed by the bank. The risk capital was only spent on projects. There was no balance sheet for the fund itself, only project accounts: so much money spent, so much received.

Like any innovator, Birman made her share of mistakes. She did not initially anticipate the rate of return required to offset the failures. At first, the fund was seeking a return of only double its original investment and did not charge any interest. Since there is neither a stock market nor a mechanism for valuing companies in Hungary, Birman could never get her money back from taking her projects public. So in practice, during the initial period, her group made investments to be paid back out of a percentage of profit, determined according to the amount of seed capital provided.

Her major success in the early phase was financing a completely natural cosmetic, known as Helia-D, developed by the Grandma Moses of Pest, from which Innovation Fund gets 30 percent of the profits. Helia-D, made from sunflower stalks, and a second product from corn husks have been wildly successful in Hungary and are available at every cosmetics store. Helia-D is now exported to Europe and has gained a toehold in the California market. A third generation of face creme and new baby cosmetics are in the blender. The former will be made from wine-making residues, the latter from pea pods.

In 1985, Birman's Innovation Fund was cut loose from the National Bank of Hungary. Renamed Innofinance, it now has to sink or swim as an independent venture-capital company, without subsidies from its parent organization. It has a balance sheet and a board of directors and pays taxes.

Today Birman's Innofinance is a sobered business. Its mission is still to support technical innovation, but it no longer waits for the mail to be delivered. It seeks out innovation. Innofinance employs a broader range of financing tools to generate earnings. It invests, it loans, it leases. An area of active support is software development. Software groups receiving support from Birman lease computers from Innofinance and are saved the front-end expense of buying computers. An investment in a project is now expected to yield a 25 percent per annum return until repaid. Birman is a catalyst in forming joint ventures and joint stock companies, hooking up inventors with manufacturers, and developing creative financing packages. Above all, she has become Miss Innovation in Hungary. One of her proudest moments came in 1986, when she entered a store in downtown Budapest and the shopkeeper looked at her quizzically. "Haven't I seen you before? Possibly. On television? Could be. Innovation? Yes. Oh, thank you. I'm so glad to meet you. Now I know what innovation is!"

Birman is no longer alone in the world of Hungarian state-funded venture capital. If imitation is the most sincere form of flattery, then Innofinance's twenty or more competitors are good indications that Birman was a pioneer in addressing real needs of capital mobility.

As part of an overall interest in Hungarian reforms, Soviet observers have noted Birman's work. She visits Moscow frequently and is on close terms with officials from the Central Committee, Ministry of Finance, and various banks. In 1985, a Soviet TV documentary entitled "Bankers from Vaci Street" detailed Birman's accomplishments.

Birman is a strong advocate of building close economic ties with the Soviets, in part because she believes that by supplying the large market to the East, Hungarian industry can better develop the economies of scale necessary to compete in the West. The drawback of the Soviet market—the ease of selling there—must be compensated for by competing in the demanding Western markets too. The largest producer of buses in the world is not General Motors. It is the company in Budapest whose Ikarus buses can be seen in every Soviet city as well as in Portland, Seattle, and Houston, which employ the economical articulated buses consisting of two coaches connected by an accordionlike joint.

Symbiosis

The challenge of any technology-oriented venture capitalist is (1) to find the right technology and (2) to find the right people to grow the business. The difficulty for Birman of identifying the right technology is compounded by the need to find technologies with export potential for earning hard currency. Hungary alone is too small a market to support a significant manufacturing business, making exports a critical criterion for the economic success of an investment. Exportability to the West is first preference, because the technology must meet world standards and revenues will be in hard currency. Therefore, contact with the Western market becomes essential in evaluating the technical merits and commercial value of Hungarian innovations.

Birman's "screen" was Andras Geszti. I first met him in 1982 when he was working at Licencia, one of the two state-owned foreign trade organizations established specifically for selling know-how and patents. After new rules encouraging more independent business activity were issued in 1981, Geszti joined a private technology trading company, which had Elizabeth Birman as its main client. Birman required someone to help her staff screen and evaluate the multitude of projects people present to the Innovation Fund for seed money. This contractual consulting arrangement helped provide the cash flow all private technology brokers need to get established.

Geszti works mainly with Western countries, speaks fluent English and German, and took an early interest in my "reconnaissance missions" to Hungary. He looks the way I imagine the hussar of by-gone days—tall, good-looking, with a shaggy black mane and a full Turkish moustache draped over his mouth—but his charger is a Renault, and his battles are fought with patents and licenses. He is one of many people now who are looking for new Hungarian technologies that have commercial potential. Someone like me, representing Western industrial interests, becomes a dowsing rod. With the capacity to get U.S. industrial reactions to technical developments in Hungary, our organization became a valuable source of commercial intelligence. Without the interaction with the United States, Hungarian technology brokers like Geszti and those in companies of other socialist countries are hard put to know what has any real value outside their own market. The U.S. market is still viewed as the place to strike it rich.

Geszti's problem is a variation on a theme familiar to any corporate researcher in America. Frequently, an idea is not taken seriously in one's own house until a respected neighbor down the street shows an interest. Japan is increasingly the international validator of technology for Americans. American interest in Prolog, the language of Compudrug expert systems software, has grown in recent years because the Japanese have selected it as the language for their fifth-generation computers. Original Russian technology for depositing single crystal diamonds on semiconductors for thermal and radiation protection was taken seriously in the United States only after the Japanese took up the Soviet work. The same phenomenon is often true in the arts. Americans paid no serious critical attention to Edgar Allen Poe and William Faulkner until the French and British hailed their literary genius.

One of my earliest technology projects in Hungary required working with Geszti for several years and involved new technology that Birman's Innovation Fund had been financing. The technology had been developed by an engineer conducting research on ways to improve engine performance for a large state company, Raba. Among the by-products of his research was a novel design for making tantalum capacitors. Capacitors act like batteries—they store energy—but can be much smaller.

Tantalum capacitors are the most reliable among the many varieties. Their uses include storage of electrical energy in computers, communications equipment, radar, and other devices in which small amounts of energy must be available. Leszlauer, the engineer-turned-entrepreneur, operated out of his basement in Gyor, halfway between Budapest and Vienna. For three years he sought, with the support of Birman's seed money, to perfect the manufacturing process and reliability of these capacitors by using a novel filament design which reduced consumption of the expensive tantalum by 80–90 percent. Aside from Birman's patience, the other consideration which kept Leszlauer's project funded was the very fact that an American company, Maxwell Laboratories, had shown a clear interest in the technology and tested some samples with encouraging results. In the end, market considerations rather than technical ones caused Maxwell to drop the project, which would have involved the company in a different end of the energy storage market than it was accustomed to.

Though this project failed to mature, it produced for me a solid working relationship with Geszti and Birman, as well as an introduction into the closely knit world of Hungarian business women.

The Birman "Mafia"

It would be misleading to call Birman a feminist. Yet she takes an unmistakable pride in the accomplishments of Hungarian women. Ágnes Cseresnyés, for example, is the person Birman considers the real pioneer of innovative banking in Hungary.

A forty-year-old, moon-faced blonde woman, Cseresnyés can flash a disarmingly girlish smile when she's not being stubborn. Appointed in 1987 managing director of the new Unicbank, itself a banking joint venture with 45 percent foreign ownership, Ágnes Cseresnyés spent the previous fifteen years as the controversial head of the legal and venture department of the State Development Bank.

The single child of a newspaperman who was liquidated in the 1950s by the Rákosi-led Stalinist wing of the Party, Cseresnyés has blended a combative, idealistic intelligence with good political connections. Never a member of the Communist party, Cseresnyés found a well-placed sponsor early in her career. At age twenty-five, she was asked by Mr. Havas, the president of the large Hungarian State Development Bank, to work in the bank's legal department, where her earlier international experience at Chemocomplex Foreign Trade Company was needed. Three years later, the head of the legal department retired. In 1978, when she was twenty-eight, the president of the State Development Bank appointed Cseresnyés head of the department, unleashing a firestorm of controversy.

In a bank the size of hers, department heads are supposed to be fifty-year-old men. Charges of blatant favoritism, insinuations of sexual involvement, and all the other imaginable sins that could be committed by a fifty-year-old bank president and a comely twenty-eight-year-old woman charged the atmosphere of the bank for years to come. President Havas, however, had the courage of his vices, real or imagined, and stood by his favorite through the ensuing controversy. As a consequence of the spotlight she was put in, Cseresnyés became determined to make a difference.

The opportunity to push for an innovative approach to her job came almost immediately as a result of changes in the law. Cseresnyés's activism produced controversy, and, again, she held the support of the president throughout the years of internal wrangling at the bank.

Prior to 1978, there were only two types of businesses: state-owned and cooperatives, mostly agricultural. Only government ministries were empowered to create new businesses. The changes in the law in 1978 permitted the spontaneous creation of new entities called "main contractor associations." These could be formed independent of ministerial action, and Cseresnyés wanted her bank to become involved. One of the first was called Medinvest. Medinvest and other "associations" were effectively consortia designed to gather under one organizational roof all the necessary expertise to manage something efficiently. Hospital construction in Hungary was shamefully slow. Medinvest was created to speed up the construction or reconstruction of hospitals. This task was accomplished by bringing together in one organization as shareholders Medicor, the main producer of medical equipment; two building contractors; one planning company; and the State Development Bank. Through having all the important actors in the plot reading from the same text, it was possible for Medinvest to renovate the large Laszlo Hospital in Budapest in two years—a project which would normally have taken ten. Similar main contractor associations, with names such as Kulturinvest and Transportinvest, were created to speed the building of schools and transportation infrastructure.

Rather than producing more rational coordination, central planning has generally had the effect of creating large vertical bureaucracies which have no natural incentives to work together and no mechanism for imposing decisions on disparate organizations. It is not unlike the problem the White

House has in persuading executive branch departments to work together in a coordinated fashion. The magnitude of the U.S. effort has itself produced two minibureaucracies called the National Security Council and the Domestic Council.

By now, Cseresnyés's legal department had been changed to the legal and venture department. In 1980, she helped push the bank into owning shares in the "Creative Youth Association," or AIE, which was formed to promote and facilitate a wide range of innovative activities. The bank took a 50 percent position in the AIE. Her desire to plough into new territory was not winning Cseresnyés friends in the tradition-bound ranks of the bank. Not only was she promoting ideas which would seemingly compromise the bank's objectivity if it were to be an investor in the same projects in which it had also placed loans; this damned upstart was changing her job description, too!

Cseresnyés was supposed to be running a legal department, not a wheeling-and-dealing new venture operation. The way one is supposed to change jobs is to rise in an orderly vertical fashion: section head, department head, managing director. If one wanted a real change, try a horizontal one into another business. Cseresnyés was changing her job in situ. And this didn't sit well with many of her fellow department heads.

Throughout, Havas stood by, defending himself against charges of favoritism, in part by pointing out that his so-called crown princess had some of the worst physical accommodations in the bank. He had not given her any special treatment in that respect. Havas argued that he was only supporting imaginative ideas and constructive change.

Ágnes Cseresnyés describes herself as an optimist and a fighter. She was offered numerous opportunities for promotion after her appointment as head of the legal and venture department, but she chose to stay on and nurture the baby she helped bring into being. For ten years, Cseresnyés and her fifteen colleagues soldiered on, breaking new ground for the National Development Bank, spawning new smaller banking institutions with specialized financing missions such as building construction, agricultural coops, high-technology projects and creating together with Birman, Novotrade, the first joint stock company in Hungary since World War II.

One example of Cseresnyés's innovative thinking was the impetus she gave to creating Revital, an Austrian-Hungarian joint venture company established to promote both restoration and new commercial and residential development in Budapest.

The restoration of the Moorish-style Dohany Street Synagogue, built in the 1850s, is the first large-scale project to be undertaken by Revital. The push behind the restoration of this three-thousand-capacity synagogue, second largest in Europe, actually came from the World Jewish Congress. Its president, Edgar Bronfmann, negotiated with the Hungarian government to persuade it to contribute 50 percent of the $30 million budgeted to rehabilitate the synagogue. The balance would be raised by the World Jewish Congress, whose $15 million campaign was launched in the Waldorf Astoria Hotel in November 1987. Revital is to be the general contractor and architectural supervisor for the huge three-year project.

In addition to managing the restoration of the Dohany Street Synagogue, Revital is considering buying the Otto Wagner Synagogue, not far from Dohany Street, which the Hungarian-Israeli Jewish Society wants to sell in order to assure its proper maintenance. Under the current thinking of Managing Director Peter Vegh, Revital would lease it to the government as a cultural center.

There is a great shortage of office space in Budapest. Revital plans to use twenty-two thousand square meters of donated real estate, part of its initial capital contribution, to build offices for Hungarian insurance companies and other businesses whose huge savings will get a better return invested in land, with prices' rising 15 percent a year, than in a savings bank.

For the inhabitants of the dilapidated housing in the Sixth

District, Revital offers a seven- to nine-month turnaround on an apartment building rehabilitation that used to take two to three years under the direction of the local city councils. One reason for this improvement is that Revital has as its construction partner the largest state construction firm in Budapest. This firm has the resources to tackle construction projects with greater efficiency than the mosaic of small contractors who would normally be used in city rehabilitation projects, which are typically unprofitable and uninteresting to the larger builders. As Revital brings access to the specialized construction equipment of its Austrian partners that is not available in Hungary and engineering know-how as well, the large state company is willing to work with Revital on all its projects, including those it would otherwise not accept.

For more affluent foreign travelers, especially Hungarian-Americans who want to divide their time between the United States and Hungary, Revital plans to build apartments privately financed on a time-sharing arrangement.

Obviously, Cseresnyés could not have dragged her bank kicking and screaming into the modern age without strong support from others with influence. One of the problems facing socialist economies which Hungary has gone furthest in addressing is the need for capital markets. Increasingly, all countries are recognizing that innovation and productivity can be furthered only if people with good ideas have ways of getting money to implement them.

Much of the Hungarian reform process has been driven by financial leaders willing to take risks and break traditional patterns of thinking. Laszlo Andor, past president of the National Bank of Hungary and others such as Lenk, Pásmándy, Demján have shown a readiness to innovate. With twenty-three commercial banks now in operation, Hungary has a modest degree of competitive banking. An incipient domestic bond market, which issued 15 million forints ($300 million) worth of bonds in 1987, is developing. "Privates" are bonds owned by private individuals, which can be issued either by state banks or by cooperatives seeking to finance projects. "Governments" are owned by state banks or state companies. "Privates" get 11 percent interest, and holders pay no taxes, whereas "governments" get 14 percent, and holders pay taxes of 50 percent. A stock market in which individuals and companies can trade in an open market is getting established and will permit 100 percent foreign ownership of some companies. Until recently, the only stocks were nontransferable shares held by employees of cooperatives and a few joint stock companies such as Novotrade in which large companies own shares. Novotrade shares were sold for 175 percent of their nominal value in 1989.

What sort of prescription for the future does Ágnes Cseresnyés see?

> First, the younger generation must get involved in the political process. The system needs their energy, flexibility, readiness to innovate. You can't appoint a sixty-year-old person to head up a large industrial organization and expect change and innovation.
>
> Second, there is too much bureaucracy in Hungary and in the world. Bureaucracy is one of the main inhibitors of innovation. Just the sheer amount of paperwork required to do anything discourages action. New organizational forms and more experimentation are needed. There has to be a way of creating smaller units within large organizations, of making it easier for small entities to be created.

Cseresnyés is not a believer in Western-style efficiency accompanied by unemployment. She believes there should be a middle ground where efficiency can be achieved without workers' suffering. In the near term, she thinks this is possible because of the great need for basic services and small shops and retail stores to provide goods; if unemployment is created as a result of the new measures, much of it can be offset by making it easy for people to become self-employed, fulfilling simple everyday service needs. Though Hungary now has private enterprise, there is still too little to offer effective competition, and she does not think that private people treat customers any better than state people. She also thinks that another way to soak up unemployment and provide new jobs is to sell off the assets of inefficient state companies and allow foreign companies to have 100 percent ownership and produce whatever they want.

Cseresnyés thinks *glasnost* is good for Hungary. In the past, conservatives and people resistant to change always wanted to know what Moscow would think. Now Moscow is on the Hungarian bandwagon.

What has motivated Ágnes Cseresnyés? "In our system, money is not an important incentive to do anything. I have worked and fought because I want to make the system work better. I am a biological optimist, but the key is not to give up. You have to keep fighting."

Another member of the "Mafia" directly involved with technological innovation is Susanna Olah, general manager of Biotechnika. Her career overlaps with Birman's beginnings in the early 1970s, when the Vepex process made such a big impression. By 1978, the Vepex process for deriving protein cheaply from grasses had attracted the attention of Swift Corporation in Chicago and Mitsubishi in Japan, as well as Russian buyers. Swift had already signed and had a pilot plant under construction, and the Russians were about to sign a contract. The technology had such potential that Vepex had received permission to set up its own commercial arm bearing the same name, vested with foreign trading rights.

Olah was selected to become general manager in 1980 only to preside over the technology's demise, or at least its interment, in a state of suspended animation. Economic conditions changed. By 1980, the price of soya, the main competitive source of animal protein, fell, and the economics of Vepex no longer looked so interesting. Olah tried to diversify the business base of Vepex, turning it into an engineering company serving the agricultural and pharmaceutical industries, but by 1983, there were few successes to show.

Today, Vepex is the commercial arm of Biotechnika, a company created in 1984 with capital from Birman's Innovation Fund, Cseresnyes's legal and venture department while she was at the State Development Bank, and the Biological Research Center in Szeged, as well as grants from OMFB, the State Office of Technology Development.

Szeged is the center of a complex of four large research institutes of the Hungarian Academy of Sciences. Each institute specializes in an area of the life sciences: the Institute of Biophysics, Institute of Genetics, Institute of Biochemistry, and Institute of Plant Physiology. In line with the growing emphasis on using local brainpower the Szeged Biological Research Center became a focus of attention of those who wanted Hungary to participate in the biotechnology revolution.

OMFB was interested in identifying research results that had commercial potential and getting them marketed, be it for Western, Soviet, or Hungarian buyers. Biotechnika was formed because the Szeged Research Center lacked both the motivation and the skills for commercial development of its own research results. Biotechnika acquires rights to new developments coming from the laboratories which it judges to have commercial potential.

As of 1988, Biotechnika was essentially an R&D firm with a portfolio of over forty projects and a staff of twenty-four scientists. Among the particularly promising new technologies that have been pursued are a high-efficiency gas-liquid aeration system, which has been sold to the company John Brown in the United Kingdom, and a monoclonal antibody for the diagnosis and treatment of chlamydia, a microorganism that produces a common vaginal infection. Biotechnika has been selected as a partner by Sovgene, one of several so-called scientific and technical complexes for coordinating all leading Soviet research efforts in areas of technology critical for the future.

In a small country such as Hungary, all paths in a particular field eventually cross. Like Darvas, Susanna Olah started her career as a chemist at EGYT Pharmaceutical Works. She prides herself on being able to get along with anybody, thanks to her diverse training at EGYT. There she worked on a variety of jobs, ranging from laboratory technician to pilot facility operator to shift worker on the factory floor. After six years, Olah became assistant to the technical director of EGYT, only to decide that she really wanted to go over to Gedeon Richter Pharmaceutical, the largest of the Hungarian pharmaceutical producers. As the five major pharmaceutical companies had a pact among themselves that they would not steal personnel from each other, Olah was told by the general manager to work somewhere else for a while. She took a job at the Ministry for Heavy Industry as a project manager in the organic chemistry section responsible for investment decisions in the pharmaceutical industry.

In 1970, Olah was one of the few female professionals in the whole ministry. This was also the miniskirt era, which she took full advantage of to the consternation and confusion of her male colleagues, who inevitably mistook her for a secretary at first meeting. This provided a regular source of amusement for Olah. Out-of-towners would enter her office and, assuming she was a receptionist, take a seat and engage in pleasantries, while wondering where the project manager they'd come to visit had gone. Clearing her voice after a certain amount of idle chit chat, Olah would ask, "Shouldn't we get down to business?" inevitably producing a look of sheepish embarrassment in the visitor.

The same advantage of surprise worked even more to Olah's benefit when she moved over to Gedeon Richter to become director of export for the hard-currency markets.

> It's only an advantage to be a woman in business, particularly if you're dealing in South American countries as I was. All the men there think they are Don Juans and have great designs when they meet a woman. A woman can take great advantage of men's self-delusions. With women, men's defenses tend to be lowered. Their protective instincts mingle with their sexual fantasies.
>
> At Gedeon Richter I had a second child. When I was six months pregnant, there was an important agreement to be negotiated in Brussels, and the General Director asked if I would go. I agreed on the condition that I could fly first class and didn't have to carry any luggage. Showing up six months pregnant alone at a business luncheon in Brussels with four men provided me with a tremendous psychological edge that helped bring the deal to a successful close.

Susanna Olah's career is a blatant challenge to women who believe they must sacrifice their femininity to succeed in the masculine world of business. No man in a dress, Olah took full advantage of her combined competence and attractiveness.

A remarkable proportion of Hungary's chemical industrial leadership is composed of women. The largest of the five major Hungarian pharmaceutical companies was run for many years by the legendary Mrs. Varga for whom Olah worked. Taurus, Hungary's largest producer of rubber products, was sluggish and thick around the waist until taken over by Ilona Tatai. A Ph.D. and chemical engineer, she revitalized the company and turned it into one of the country's better managed and more innovative enterprises, now operating a wholly owned subsidiary in New Jersey. Tatai also serves on the Politburo of the Hungarian Communist party. The general director of the largest producer of products made from polyvinyl chloride (PVC), Hungarian Chemical Works, is the dynamic Elizabeth Feher, also a chemist. Birman and Cseresnyes are two of several women who have obtained influential positions in banking circles. Though few in number, these Hungarian women have achieved significant positions through a combination of good technical training, sheer ability, and connections.

Resistance to Change

Although viewed by the West and by many in the East as progressive, Hungary has not found change easy. In the final Kádár years, opposition, natural inertia, and exhaustion among the advocates of reform seemed to be conspiring to produce economic stagnation and indecisiveness. Kádár's successor, Károly Grósz, has breathed fresh life into the movement to drive Hungary further toward a market style of socialism, abetted by Gorbachev's own activism in the same direction.

Valuable insight into the sociology of change was provided by Eva Sitai, a colleague of Geszti who introduced me to András Hegedüs. Hegedüs had been an economist at the Research Institute for the Plastics Industry for several years before drifting into sociology. As an economist at a research institute, he was supposed to tug occasionally at the sleeves of the research managers to suggest that considerations of cost and benefit be applied to evaluating the research tasks at hand. In the days of technocratic supremacy, the very idea of applying economic criteria to the solution of technical problems was unheard-of. Disillusioned with being a token presence in the institute, Hegedüs moved over to the Academy of Sciences.

There, he and a colleague, Gyula Kozak, interviewed 156 senior executives from 1980 to 1982 as part of a project they carried out for the Institute of Economics of the Hungarian Academy of Sciences. The interviews produced twenty-five thousand pages of transcripts, creating the most massive record of oral history research in postwar Hungary. The purpose of the study was to determine how executives rose to the top, what motivated them, and how they viewed the economic impasse in Hungary at that time. Eager to compare notes, I settled into an evening of discussion at his dingy, fin de siècle apartment with Eva, fueled by a large plate of oozing Hungarian-style napoleons and brandy.

Before getting into details, I was provided a review of postwar history: the years 1945 through 1948 constituted the "democratic" period—although András and Eva finally agreed that it was really only democratic in comparison to what followed. There was a pluralistic political system during that period, but it existed under the shadow of Soviet military occupation. In 1946, the big banks and largest factories were nationalized. Key business leaders lost their jobs. This was also a time when people were literally grabbed off the streets, some of them recently returned from prison camps, to go to new camps for "*malenkaya robota*," or "a little work," as the Russian soldiers cutely put it. The Hungarians had been unenthusiastic allies of the Nazis, making them targets for in-kind reparations.

In 1948, the first elections produced a 17 percent vote for the Communists. Three years later, the vote was 100 percent Communist. Arrests, intimidation, harassment, and selective liquidation by the better organized Communist party, supported by the Soviet occupiers, resulted in the quiet elimination of political opposition. Though Americans would call it expropriation, the land and wealth were, in Eva's eyes, democratized during this period. The large agricultural estates of the nobility were broken up and redistributed to the peasants. Educational, medical, and cultural opportunities were made more available to the poor.

Under the leadership of Mátyas Rákosi, a period of industrial restructuring along Soviet lines ensued. The result was an economic disaster. The Hungarian economy was converted from one specializing in light industry to the heavy-industrial, metal-eating model of the Soviets, making machines for the sake of making machines. Steel was king. Consumption was squeezed to near-extinction.

In 1949, the show trial of Laszlo Rajk brought about the final elimination of Rakosi's lingering political enemies, consolidating the power of his Stalinist wing of the Party. During this time, 1948 through 1956, most of today's managers of Hungarian industry were installed: Party administrators whose chief qualification was Party loyalty or proper class origin. They are people whose spine was broken by the terror of the 1950s and for whom survival meant carrying out orders, adapting, and hanging on to their positions.

Power in this system comes from position, not money. When someone loses his position, he loses everything. That is why people stay on for so long. Pensions are low. The privileges of power go with office, not wealth.

Even after 1956, only a few of the most tyrannical Stalinists disappeared. For the most part, those who were managers before the uprising stayed on in their positions. Today, Hungary suffers from paralysis. The reformers are running out of steam. The process has been going on gradually since 1968, and people are exhausted from fighting. Few actually oppose reforms, but Hungary is in an economic crisis. The country has a terrible debt, prices are going up, and people are worried. If Hungary has to reschedule its debt, prices will rise even more. Hungary hasn't the real discipline of the marketplace, yet it also lacks the discipline of central control.

The attitude toward making money is very negative. People who make a lot of money are suspect. Power and money today are negative values. There is no respect for those in power. Money is still viewed as bad by most people. Hungary has promoted an egalitarian economic ethic for forty years and it can't be easily reversed without risk.

A summary of the executives' attitudes toward the reforms was published in 1987 in the *New Hungarian Quarterly*. They fell into three broad categories: anti, pro, and weathervane watchers.

The Antis

The 20 percent who were "anti" rarely declared flatly that they were against the reforms, but rather stressed the negative side. Executives who were unenthusiastic from the very beginning tended to be older and to have come from the large state-owned industrial organizations. These were managers who had risen to the top by following orders under the system of central directives. Their intellectual energy and careers were tied up with the system that preceded 1968. More than defense of the status quo was involved: This group often held that competition and profit-mindedness harm the businesses they manage.

Concepts of worker participation in decision making were not well received by the antireformers. Ironically, some of these socialist "conservatives" almost sound like Western individualists. "There is no such thing as collective wisdom," said "A" during an interview. "A decision must always be made by one man."

On the subject of worker management relations, these "conservatives" could sound like Ross Perot. "There are no inherent conflicts, only ones that are created by poor leadership," according to one of those interviewed. Tomas Bata, the great entrepreneurial Czech empire builder of the 1930s, also believed that the struggle between employer and employee was totally unnecessary. These same people do not believe that conflict of group interests serves the good of society as a whole.

Closely allied to the views of the antireform industrial managers were those of the engineers. Their opposition was not based on principle or economic grounds, but solely on the threat they saw to their sector or "technical interests." The chemical engineer and metallurgist were each committed to the technical progress of his specialty, often associated with cheap money and plentiful investment and R&D grants from the ministries. Lacking an ideological point of view, the technocrat had the sole desire to be left in peace to meet his targets.

Weathervaners

The weathervaners supported the reforms because that was policy. Hegedüs called certain of these "declamatory reformers": people who were accustomed to adapting their rhetoric to the policy of the day. Declamatory reformers were frequently people who had risen to their position as a result of being disciplined followers of the Party line. Political loyalty, rather than expertise, was their chief currency. This group, comprising about 10 percent of those interviewed, occupied high positions, were opinion leaders, and were expected to support the reforms. Yet in the course of six to ten hours of interviews, it was clear that these same people, the older ones especially, had been conditioned by the centrally controlled system of earlier days and wished for its return. A sixty-year-old general manager summed up his credo: "A little compromise, a bit of the human element, a little bullying to the right, to the left, downwards and upwards as well. One must not deviate from the main line, but one may have private opinions too. Of course, one doesn't have to express them all the time."

A closely related breed are those governed by ambition. Their sense of responsibility is only to say what is expected, disregarding any conflicting convictions. The true weathervaner is a chameleon with no views other than those that serve the goal of self-promotion. Motivated less by a sense of duty and undisturbed by conflicts of conscience, this type of "reformer" is driven by the hunger for power. A rising young star in his forties openly stated his philosophy as follows: "In order to have the career I longed for as a young man, I always need to recognize what it is I must comply with. I carefully examine my chances of obtaining authority and honors and think that by the age of fifty I will hold a pretty high office. To assure only steady but not too rapid advance, I need to avoid conflict with one and all, to move smartly and to know what they expect of me."

The Pros

Finally, there were the out-and-out true believers in reform, who constituted 40 percent of those interviewed. The committed reformers generally are younger, better educated, and heavily represented in the agricultural sector. Agricultural managers believe in reform. They attribute the success of the agricultural sector to the reforms and the success of the reform process to the improvements in agricultural productivity.

Hegedüs explained to me why this phenomenon occurred in agriculture. Although not generally referred to as such, a strong agricultural lobby developed after the disastrous policy of heavy industrialization in the 1950s. Chief "lobbyist" was Imre Nagy, premier from 1953 to 1955. Executed for his role in the 1956 uprising, he had spoken out for agricultural reform and for the privatization of small-scale retail businesses. Nagy thought that shopkeepers, barbers, shoe repairmen, and similar species should be independent from state control. Although he lost the battle for the small businessman, Nagy won allies in his fight to make agriculture more effective. With shrewd insight, the Party recognized the importance of food and began to allow principles of competence and consensus to creep into the organization of agriculture. Unlike industry, which operated as a military command structure in which directors followed orders and gave orders, agriculture became relatively more democratic, at least formally.

District councils are vital "organs" in the Party's supervision of local agriculture. Though the Party nominated the directors for whom the local agricultural cooperatives would cast votes, efforts were made to ascertain in advance that there were at least no strong objections to the candidates. Partly because of this consultative approach with the cooperatives and because the Party wanted agriculture to be efficient, expertise started to win out over politics earlier in agriculture than in other spheres of Hungarian economic life. By the 1960s, prewar agricultural experts were returning to important posts.

In parallel, an agroindustrial complex grew up. Powerful industrial interests with a stake in chemical fertilizers and agricultural machinery began to ally with the farming interests. Political heavyweights such as Andor Huszár of TVK Chemical Company and Ede Horváth of Rába, a large farm implement enterprise, often sided with the farming interests.

In the minds of people living in the country, the reforms are closely linked with the perceptible improvement in the standard of living. Career mobility also became greater in agriculture than in most other sectors of economic life. Graduates of agricultural schools with advanced training possess a versatility that enables them easily to fit a variety of institutional positions—president of a cooperative, head of a county council, government ministry.

As reported in the *New Hungarian Quarterly*, the career of "Zoltan," the president of a cooperative farm, who is in his late fifties, is illustrative. He was born the seventh child of a poor peasant family. In 1942, Zoltan began studies at an agricultural college but was interrupted by the war. In 1945, the land distribution program of the Communists attracted him to the Party, though he never abandoned his strong religious convictions. At age eighteen, he began work at the county council, which was engaged in enforcing collectivization, and by twenty was wielding great power over agricultural affairs in the county. Religion need not be an issue, according to Hegedüs, even for a Party member, as long as one practices it quietly.

After severe disagreements with his superiors at the council, Zoltan left to become chairman of a poorly performing farm cooperative. Though he made the co-op prosperous within a short time, he was, nevertheless, an "outsider" to the village and was defeated in the first reelection vote. Zoltan moved on to become a wine buyer for a large state company, then was selected in 1960 to be the chairman of another cooperative in the wake of a new wave of farming reorganization. Again clashes over policy with county executives followed, then another job at another state company. Finally Zoltan became chairman of yet another failing cooperative. This time, when he finally turned the co-op around as a result of completely changing its production, he had also achieved international success with a new technology he developed and traveled all over the world selling it. Such is the roller coaster of fortune on the Puszta.

A significant number of supporters of the reform are in industry. These people believe strongly that the efficiency of the economy is tied closely to ending the producers' mentality of dependence on their ministerial superiors. The reformers want production enterprises to make their own decisions about the products they make, where they sell them, and the prices they charge. Equally important, reformers also believe, as do antireformers but from a different perspective, that predictability and equality in the rules of the game are essential. These are people who believe enterprises must take risks and be prepared to sacrifice short-term results for long-term gain, but that the conditions of risk must be equally distributed throughout the economy.

Generation Gap

One of the most important divisions in the group of 156 was age. Two-thirds were over fifty. Characteristic of those in their fifties and sixties were careers based on a strong political orientation and a technical education. They had witnessed the consequences of being politically "wrong" during the Stalinist period.

Typical of the older generation of successful managers who rose to the top in the 1950s was "Béla." Béla was born in 1908 into a large working-class family and experienced the white terror of 1920 after the fall of Béla Kun. It was a struggle for his parents to pay for his secondary education. In the thirties, he joined the illegal Communist party and deserted the army during the war to fight against Hitler. In 1945, the Party nominated him to head the factory committee of a large chemical plant where he had worked earlier as an electrician. This appointment was the beginning of his executive career. Though "Béla" knew little about chemical engineering, he knew a lot about the organization of the plant and its people. More important than his technical knowledge, or lack of it, was his intimate familiarity with the people and their problems. The experienced electrician, sensitive to the "human side" of management, could handle workers of all stripes better than more highly trained executives who had never gotten their hands dirty. Though he never mastered chemical engineering, "Béla" did complete a two-year advanced course in economics and technology at the Red Academy at forty-three.

Retired at age seventy-seven, "Béla" recalled of that period, "Catching up with what one missed as a boy is almost impossible. They taught me to think, so I could see things in a different light, and to realize that brain work was the most difficult of all activities. Yet one could not really get much practical knowledge at the school."

Successful executives in their forties are still uncommon, as there are no incentives for executives in their sixties to take early retirement. The generation in their forties differ markedly from the older executives. Their higher education was generally freer of dogma than their predecessors', and they have not known personally the traumas of the Stalinist era. In contrast to the older managers, those in their forties have many economists in their ranks, reflecting the growing recognition that management of the economy or of state companies requires an appreciation of the way scarce resources are allocated. The class background of the younger executives is also different. Managers in the immediate postwar period were from the lower classes and often had memories of severe poverty. These are people who consider throwing away a piece of bread immoral. Today's younger managers frequently come from the old middle class and do not know the meaning of hunger.

Hungarian managers and business executives as a group are not regarded highly in the Hungarian public mind, according to Hegedus. But he believes that today's managers are better qualified than either the man in the street or economists realize. Nor do they shun taking on all the headaches of responsibility for meager financial rewards. The majority, Hegedus believes, simply are passionately engaged in their jobs. They like coping with the challenges and are obsessed by them, and many, like Ágnes Cseresnyés, are motivated by the plain stubborn desire to make the system work better.

Since I talked with Hegedus, two things have changed the situation, causing the Party leadership to move decisively toward more market discipline. One is the worsening economic crisis and the other is *glasnost*. *Glasnost* has made economic experimentation in the direction of freer markets more respectable. The government started in January 1988 taking painful measures to reduce its outlay for subsidizing inefficient producers. Twenty-five percent of the 600-billion-forint ($15-billion) state budget supports failing industry.

State companies will be completely freed of ministerial control and will have to sink or swim. To support subsidies, consumers are now paying a 20 percent value-added tax and a 60 percent income tax on personal income over twenty thousand dollars per year, though this will initially be paid by the employer. The 60 percent rate affects relatively few people, but there is uneasiness everywhere about the possibility of unrest. The Party leadership has wisely provided a safety valve for the trouble which might result from the greater economic insecurity. Freedom to travel is now allowed anybody with a minimum of sixty dollars in hard currency. There is a virtual open border between Austria and Hungary now that visas are no longer required.

The Soviets have paid close attention in the past to Hungarian reforms and will, no doubt, continue to do so. There is now an Innovation Bank in Leningrad, and private cooperatives of all kinds are springing up in the Soviet Union to provide software, produce scientific instruments, provide computer modeling for training shuttle pilots, sell business consulting services, and operate restaurants, to name but a few varieties. Freedom to travel presents a crucial litmus test of political self-confidence. The Hungarian experience, which now includes official recognition of non-Communist-controlled political parties, will certainly guide the Soviet leadership as it, too, considers further freedoms.

Chapter 2

The Wizard of Prague

The surest way to corrupt a young man is to teach him to esteem more highly those who think alike than those who think differently.

FRIEDRICH NIETZSCHE

Nineteen sixty-eight, the year Soviet tanks moved into Czechoslovakia to crush the Prague Spring, ironically is the same year Hungary began the two-decade-long reform process that has now made it the consumer paradise of Eastern Europe. An important intellectual antecedent to the failed Czech and successful Hungarian reform process was the work of Czech Ota Šik. An electrician turned political economist from Pilsen, Šik became a leading theoretical architect of market socialism in pre-1968 Czechoslovakia. As early as 1958, when Czechoslovakia was undergoing its first post-Stalinist reorganization, Šik was thinking about the limitations of a socialism that ignored the function of prices and relied on oversimplified quantitative approaches to production planning.

In his important work, *Plan and Market under Socialism*,

first published in 1964, Šik also criticized the administrative and managements side of existing socialism, which did not recognize the potential for arousing and utilizing people's initiative. There was nothing inherent in socialism which should deny incentives for exercising initiative, but it was still grounded more on revolutionary will and dogmatism than on analysis and real economic findings. These ideas of Šik's provided much of the intellectual underpinnings of the Czech reform process in the 1960s.

As director of the Institute of Economy of the Czechoslovak Academy of Sciences, Central Committee member, and later deputy prime minister under Alexander Dubček, Šik was inextricably bound with the discredited reformists. A compromised political economist with a vision of a third way in ruins, he had the alternative of fleeing or living in internal exile.

Šik lives and teaches today in St. Gallen, Switzerland. Another Czech free spirit of that same period who identified himself with Dubček reformists was Otto Wichterle. One of Czechoslovakia's wealthiest and most respected personalities, he lives happily in Prague and has a summer home in upper Moravia. His career illustrates the degree to which independent-mindedness can be tolerated by the system, particularly when combined with "hard" scientific contributions which can benefit the national economy in tangible ways. Otto Wichterle is a chemist of world rank who fathered the soft contact lens industry in the United States, a man who lists foreign trade as a hobby on his résumé, a political "enfant terrible" at home, a former two-thousand-dollar-a-day consultant to Bausch and Lomb, owner of one of the first hard-currency bank accounts in Czechoslovakia, and a man honored in his native Moravian village for fixing the clock on the church's steeple.

Path to the Institute

"A measure of a man is how he deals with those unexpected confrontations with reality called chance," commented Wichterle after I had noted that his career seemed to have been directed by quite unplanned events.

The first was shortly before Easter 1956. Wichterle was attending a large meeting of educators and university professors in Prague's Fučik Hall organized by the Party to discuss educational policy. The conference was open to all. Wichterle was walking with Professor Francis Šorm, the president of the Academy of Sciences, when they ran into the prime minister, Viliam Široky. Široky had never met Wichterle, but he knew him by reputation. Just then, he had under his arm a study prepared by Wichterle comparing his own process for making caprolactam to a Russian one which was under consideration by Czech industry.

Caprolactam is a chemical intermediate which is different from that developed by Du Pont in the 1930s to produce nylon. Wichterle and a group at the Bata Chemical Research Institute in Žlin had developed the technology for making a new kind of nylon during World War II in parallel with German researchers at I. G. Farben. Wichterle argued that his continuous process was more economic than the German batch process which the Russians had adopted. His analysis was part of the document in Široky's hands.

When asked politely by Široky how he was doing, Wichterle indicated that he was doing just fine, except that he would be a lot happier if he had a passport and were allowed to travel. He very much wanted to attend an important conference in Rehovoth, Israel, the following week. Sponsored by the International Union of Pure and Applied Chemistry (IUPAC), it was the first international conference on polymer chemistry, and many chemists of world stature would be

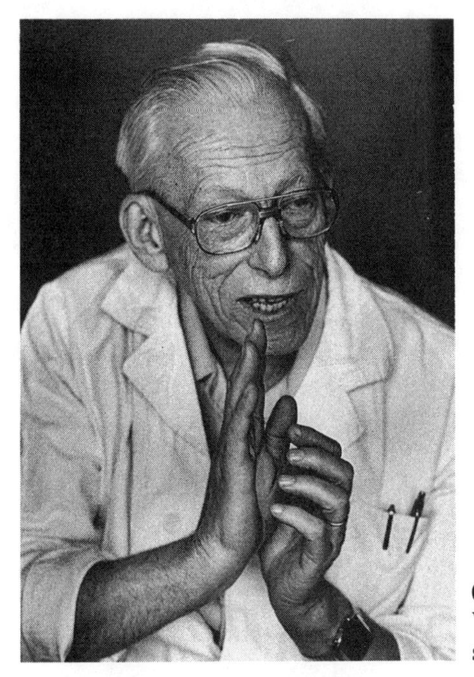

Czech chemist Otto Wichterle, creator of the soft contact lens.

there. Široky expressed surprise that Wichterle did not have a passport, not knowing that he had been deprived of his since 1948. Coming from a well-to-do bourgeois industrial family, Wichterle had a class background that was not considered suitable for Party membership or travel privileges. Throughout his whole distinguished career, he was never asked to join the Party.

Široky said he'd look into the matter and clear it up. Wichterle thanked him, but left doubting that anything would come of the prime minister's gesture. The following weekend was Easter. Wichterle and his family were visiting Bečhyne Castle in southern Bohemia when a stranger approached him and presented a telegram. Surprised that anyone would know where he was, Wichterle opened the telegram to learn that he was expected back in Prague immediately. His ticket and documents had been prepared, and he was to travel to Israel the next day.

When Wichterle arrived in Rehovoth, he discovered that a clerical error had been made in computing the exchange rate between Czech crowns and Israeli pounds with the result that he was extremely short of cash, so short that he had to go to the Czech embassy in Tel Aviv for an infusion.

At the conference, Wichterle saw Herman Mark, one of the big names in polymer chemistry. Mark was a Viennese professor who had left for the United States in 1939 and had set up a world-renowned polymer research center at Brooklyn Polytechnic. Mark had casually mentioned to Wichterle that IUPAC should have its next conference in Prague. Wichterle casually said yes; it would be nice to do that. At one of the following sessions, the subject of the next year's meeting came up. Mark volunteered that Wichterle had invited the group to have its next meeting in Prague. Wichterle, of course, had no authority to make such a decision; that had to be made by the government.

Thanks to the mistakes over the exchange rate, Wichterle now had contacts at the Czech embassy who orchestrated the high-speed minuet required for Wichterle to give the IUPAC committee a formal reply before it disbanded. Phone calls were made to the president of the Academy of Sciences asking approval. Then the president of the Academy, Šorm, called the government. Within twenty-four hours, they approved the invitation.

There was nervousness in the government about hosting an international conference. This had not occurred before, at least in chemistry. How would Czech scientists look in comparison with foreign ones? Would it be organized efficiently?

Wichterle was named head of the committee which organized the conference a year later. They trained thirty-six translators in French, Russian, English, and German. Three Nobel Prize winners came: the Russian, Semenov; Natta from Italy; Ziegler of polyethylene fame from West Germany; and many other distinguished names. The conference came off like a choreographed stage performance with the Czechs' heavily represented among the scientific presentations. It was a major political event and brought forth a delegation from the Central Committee of the Party to observe the goings-on. The Czechs acquitted themselves well scientifically and organizationally. The conference was judged a great success. It gave considerable impetus to the decision of the Party to create a new institute dedicated to polymer chemistry.

The next year, a second, more important unplanned event took place. Wichterle was fired from his post at the chemical faculty of the Technical University in Prague. His dismissal resulted from a fundamental disagreement with his boss, the dean of the faculty, over the manner in which the chemical curriculum should be organized. At the time, the curriculum was divided along lines of industrial categories: a ceramics department, a synthetic rubber department, and so forth. The organization was modeled on that of the Mendeleev Institute in Moscow.

However, the mission of the Mendeleev Institute was different from that which Wichterle had in mind. In Russia, where severe trained manpower shortages existed, the point of the Mendeleev Institute was to produce well-trained technicians and managers who could walk into a synthetic rubber plant and have a working familiarity with all aspects of the technology of that industry. Functioning effectively in a factory environment required superficial knowledge of many fields rather than deep scientific knowledge.

Wichterle's goal was to pursue basic research more suitable for a university. He wanted to organize the curriculum along the lines of basic academic disciplines, such as mathematics, physics, physical chemistry, and organic chemistry, which could be applied to the investigation of a broad range of research topics. In Wichterle's mind, there are no macromolecular chemists, which is to say, polymer chemists. It is a subject for interdisciplinary research best performed by scientists representing different theoretical disciplines.

It was over basic pedagogical philosophy that Wichterle dug

in his heels against higher authority. Though he lost the battle, he won the war. His dismissal led him to have a chat with Havlin, the Party's man responsible for science and education policy and himself a chemist. Havlin told Wichterle that a Party decision had been made to build a new institute for macromolecular chemistry within the Czechoslovak Academy of Sciences. Would he like to be director? Wichterle seized the opportunity to put into place the philosophy which cost him his job at the Technical University.

But before doing so, he had the more down-to-earth challenge of getting his new institute built and equipped. Again, unexpected events helped Wichterle.

In 1960, the Ministry of Culture asked the president of the Academy of Sciences, Professor Šorm, what research of practical value had been produced since the academy's inception in 1952. Wichterle had just been appointed director of a yet-to-be-built new Institute of Macromolecular Chemistry, when Šorm asked him to provide a demonstration of something practical coming from the Academy. Wichterle's past work with caprolactam was to be Exhibit A. Caprolactam is a ring-forming molecule that can be used to make nylon, similar to the nylon Du Pont scientists discovered in the 1930s. The Du Pont nylon from which women's stockings were made came from a different and more expensive chemical raw material, though its final properties are virtually the same as those of nylon produced by caprolactam.

In an effort to find something useful to do in the laissez-faire research atmosphere at Bata Chemical Research during World War II, Wichterle joined a team that was trying to reproduce the results of Carothers, the Du Pont scientist who invented nylon. In the course of this, Wichterle came across a paper by Carothers which described his unsuccessful efforts to make nylon fiber from caprolactam, later to be known as Nylon 6. Carothers thought the inability to draw fibers was due to the determination of caprolactam molecules to form large rings instead of long chains. After futile attempts to prevent the formation of these rings, using a high-voltage electrical field, Wichterle discovered that the problem lay in the poor-quality starting material Carothers was using. Pure caprolactam produced fiber-forming polymers easily. This early work of Wichterle's was completed after the war while he was at the chemical faculty of the Technical University, where he also learned to make very large solid castings from Nylon 6. As nylon has good abrasion-resistance characteristics, bearings and other machine parts became an obvious application.

The demonstration took place in the office of Prime Minister Široky, where thirty other ministers, deputy ministers, and top-level officials gathered around the conference table. Like a modern-day Merlin, Wichterle had his beakers and potions on a side table. The experiment was a simple one, lasting five minutes. Two beakers of preheated molten caprolactam, each with different components of the catalytic system needed for polymerization, were mixed together in a jar. The jar was connected to a temperature sensor and millivolt meter which would allow the onlookers to follow the changes taking place.

Wichterle explained in advance exactly what was to be expected. A spontaneous rise in temperature would occur during the time the transformation of the separate ring molecules into a long chain was taking place at a predictable rate. This was represented by a picture of a conversion curve provided on a handout to all present. The conversion curve shows the relationship between time and percentage conversion of the monomer building blocks into polymerized chains. At full polymerization, the original liquid mass would become crystalline and shrink, causing the block to fall out of the jar.

Thus informed, the maestro began his work. The liquids were mixed, and the millivolt meter began to move as predicted. As the temperature in the jar increased, the material became denser. Finally, in a couple of minutes, full polymerization would occur. Wichterle took the jar over to the ministers' table and began to count down 5, 4, 3, 2, 1—0. Exactly on schedule, a round, steaming hot block of nylon dropped to the table.

If a camera had been concealed in a wall, posterity might have been able to witness what must have been the first and only game of hot potato played in high government chambers. As the block began to roll around the table, those seated at the edge put out their hands to prevent it from rolling off onto the floor, toasting their digits in the process. Laughing, crying, shouting, and blowing on their fingers, the thirty-odd officials finally wrestled the nylon plug under control to hear the rest of the story.

Wichterle then presented his economic analysis, demonstrating the expected savings to the economy. These were considerable, mainly because the caprolactam nylon would be so strong and abrasion-resistant that it could be used as a substitute for many metal components requiring the importation of expensive alloys. Nylon parts would not rust; they would be lighter and have a very low coefficient of friction.

After hearing the benefits to the Czech economy from the longer service life and materials saving to be gained from the use of Nylon 6, the minister of the Machine Industry asked Wichterle, "You've shown us what you can do for us, what can we do for you?" Unprepared for this question, Wichterle said that he did not need crowns. However, he did need hard currency, to buy the most up-to-date research equipment available, since polymer science requires precise measurement of the different properties which characterize a polymer, such as visco-elasticity, shape, density, and distribution of molecular weight. He asked for $500,000 and received it forthwith.

This sum helped considerably with what was needed for the inside of his institute. Despite the high priority given to the construction of the building, Wichterle still had to contend with the everyday work habits of the Czech laborers.

Eager to get all the external work finished before winter set in, Wichterle chided, cajoled, and ultimately bribed his construction gang with a case of liquor. Telling the laborers that in the Bata days it was expected that a crew would complete one floor per week, Wichterle dangled twenty-four bottles of rum before their thirsty eyes, as reward for completion. The floor was finished in twenty-four hours. Later, when an extra stimulus was needed to hook up some outside utility lines to the building, Wichterle offered the gypsy work crew hard cash of five hundred crowns, or one week's wages, to complete the job.

Wichterle was paying these "bonuses" out of his own pocket, but word still percolated up to the Party officialdom of the new director's unorthodox methods of getting the job done. A visit by a representative from the "higher organs" let it be known that such methods had to stop. Wichterle was accused of corrupting the workers and threatening the whole system of socialist inefficiency.

Duly slapped on the wrist, Wichterle went on to make the institute world-famous in polymer circles. The Institute of Macromolecular Chemistry patents more research, sells more licenses, and produces more income than any other research institute in Czechoslovakia. Scientists from all over the world visit it.

New organizations generate energy and enthusiasm if properly led. When Wichterle was named director of the institute in 1959, he was already a chemist of considerable stature, having won Czechoslovakia's highest scientific award, the State Prize for Chemistry. He had authored a highly popular textbook on inorganic chemistry, even though trained as an organic chemist. He had a reputation for independence and attracted to the institute many of his best students from the Technical University. Influenced by the way his friend Hermann Mark organized polymer research at Brooklyn Polytechnic, Wichterle also had the blueprint of the way polymer research should be organized in his head.

Polymer research is inherently interdisciplinary, made easier when all the branches are under one roof. One of the afflictions of other polymer research centers around the world was the fragmentation of disciplines brought on not only by physical separation but by the widespread professional antagonism which readily springs up between physicists and chemists. Physicists are prone to look down at chemists as "cooks." But physicists cannot do good work without chemically well defined starting materials. Chemists also need physicists and their sophisticated equipment to define the different properties of polymers, the versatile plastics that can be "engineered" to be as soft as a contact lens or as hard as a flak jacket.

To combat this tendency, from the outset Wichterle established the parity of chemical and physical contributions to any research result. No one was allowed to be put in the role of doing "service" work for the other. All who contributed to a paper were considered authors. There were no footnotes saying the synthesis work was done by so and so, or analytical services were provided by such and such. Chemists and physicists were made to realize they were equally important and equally interdependent.

Wichterle's success in stimulating a productive and cooperative research environment came not only from moderating disciplinary schisms endemic to the field but from other principles as well. He was a test tube sniffer. A research director must also do research, be in touch with ongoing activity, visit the laboratories. These were basics for Wichterle.

Wichterle's penchant for experimentalism was accompanied by a passionate dislike for the artificial distinction between pure and applied research. It was not a distinction he allowed at the Institute for Macromolecular Chemistry, believing that a certain percentage of people will naturally want to see something useful come of their work. This belief led to the principle of spontaneity, producing a free-floating environment in which specialists from different departments were invited to form their own spontaneous working teams around broad research problems, such as relationships between structures of polymers and their manifold properties of shape, density, elasticity, and molecular orientation.

Research planning was governed by serendipity; annual plans were based on promising results of the previous year. By allowing the "usefulness" instinct of his researchers to run its course and giving researchers freedom to define their own approaches, the Institute found its own level of applied research at about 10 to 20 percent of the total annual budget.

First Contact

While Wichterle was transforming his airy new well-equipped glass and granite four hundred–strong research center on Heyrovsky Square into an international force in polymer science, he had been unofficially carrying out since 1961 manufacturing experiments at home to produce the "eye wear" that would revolutionize the industry. This oddity was the consequence of the original mandate of the Czechoslovak Academy of Sciences, founded on the Soviet model, to pursue only pure research. Applied research was the province of industry. Though the distinction never held up in practice, in the early years, applied work was frowned upon by academy officials.

Unknown to Wichterle, rumors of his research were filtering out to the Western ophthalmological community. In the early 1960s, two New York lawyers, Martin Pollak and Jerry Feldman, had set up a business to buy patent rights for corporate inventions that had never been commercialized. In industrial parlance, these were "on the shelf." The two businessmen soon realized that unused patents were still shelved for good reasons, and there was no business opportunity in these unused corporate research results. In 1963, Pollak had the idea of looking at Russia as a possible source of new technology. In 1964, Feldman met Bob Hope's ophthalmologist at a Los Angeles cocktail party. From him, Feldman learned there were reports from Czechoslovakia of a breakthrough in polymer chemistry which solved some well known problems associated with developing a more comfortable contact lens material. In contrast to a hard lens, a soft lens needs a critical combination of properties. The material has to be tissue-compatible, to breathe, to be capable of sterilization without chemically breaking down, and to have unique optical properties.

Pollak received Feldman's telex about the rumored soft lens breakthrough while shopping for technology in Moscow; this led him to make an unscheduled detour to Prague on the way to New York. Foreign trade in Czechoslovakia is conducted by various state-owned companies under the jurisdiction of the Ministry of Foreign Trade. Polytechna is one. It is different from other foreign trade companies in Czechoslovakia in that it does not buy or sell anything tangible. It only sells rights: rights to use patents and trade secrets developed by Czech scientists and engineers. Polytechna acts as the exclusive commercial agent for the whole of Czechoslovakia; all foreign buyers and sellers of technology must deal with Polytechna in order to conduct a commercial transaction.

When Pollak arrived at the cramped, dingy offices of Polytechna, he discovered that no one there had the slightest awareness of any soft lens technology in Czechoslovakia. Not easily put off, Pollak suggested to his hosts that they call some of the institutes that might be associated with this type of work. The advantage of looking for an unknown researcher working on an unheard-of topic in a country the size of New York state is that such a request is not considered totally absurd. Prodding and cajoling, Pollak had phone calls placed to local research organizations until Polytechna finally triangulated the nearby Institute for Macromolecular Chemistry of the Czechoslovak Academy of Sciences, a mere twenty minutes away.

By the time Pollak arrived in Prague, Wichterle had moved into his second generation of home manufacturing technology. He had used such high-technology componentry as his son's erector set as a housing for homemade molds spun on flywheels powered by another son's bike generator. This was the same centrifugal spin-casting technology he had put into practice on Christmas Day 1961 and called the "Christmas apparatus," except that now Wichterle had fifteen spin casters going at once.

With samples made from his home manufacturing kit, Wichterle went to meet Pollak at Polytechna. In response to a question about the strength and durability of the lenses,

Wichterle's "Christmas apparatus."

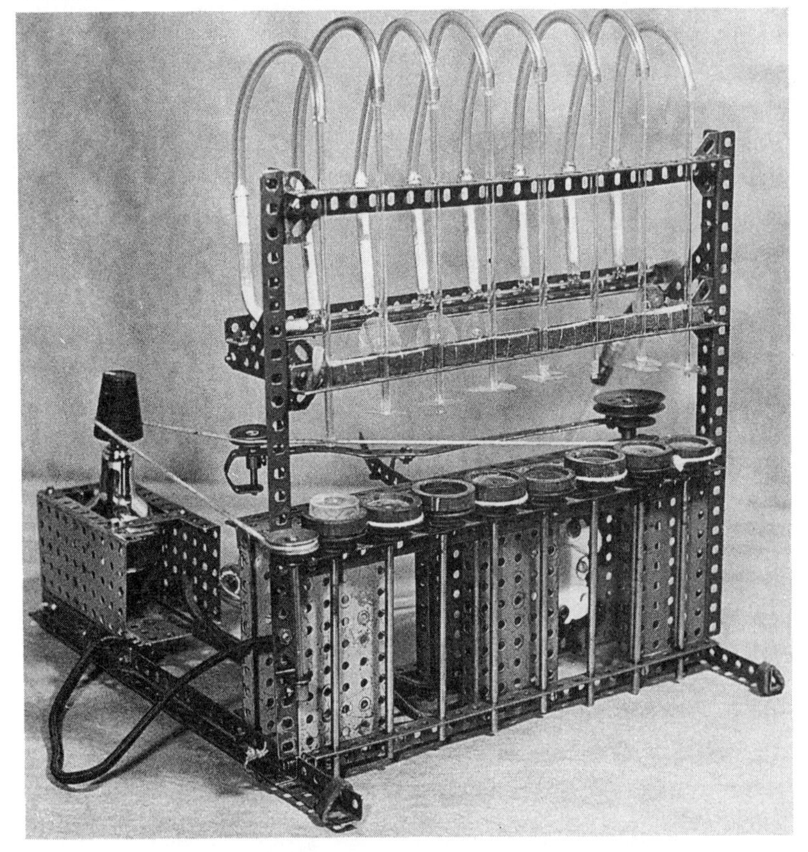

Wichterle dropped one on the parquet floor and ground it with his foot, then cleaned it off in his mouth and inserted it in his eye.

Wichterle's rough-and-ready demonstration of the physical toughness of the new soft lens motivated Pollak to round up twenty-five thousand dollars from Bob Morrison, a businessman from Harrisburg, Pennsylvania, in order to buy an exclusive option to the patent rights and know-how for producing soft lenses. With the expert advice of an optometrist from Buffalo and Morrison's money, Feldman and Pollak's company, National Patent Development Corporation (NPDC), spent the next year doing strenuous missionary work. The conventional wisdom was all against them. The optical properties could not possibly be right; they'd ruin the eye; there would be infections, ad infinitum. Like modern-day Galileos trying to persuade the sages to look through the telescope, Pollak and Feldman could not induce Professor Salvatore, a leading ophthalmology authority, to put a sample lens into his eye just to see for himself. "It can't work! Read my book!" he said repeatedly to Pollak.

None of the obvious candidates for buying the technology was biting. Pollak and Feldman had wrongly assumed that producers of hard lenses would be interested in soft lenses as an improvement over their current products. They had the distribution and marketing arranged and needed only a new and better product to advertise to their existing customers. Frustrated, Pollak and Feldman decided to approach Bausch and Lomb, a company which knew something about optics, had a solid reputation, but had no vested interest in the old hard lens technology. Pollak and Feldman were right this time. B&L was looking for new business opportunities, as their traditional precision-optics business was taking a beating from European competition.

In 1967, the vice-president of Bausch and Lomb, Dan Shuman, went to the Macromolecular Institute, where he saw the third-generation spin caster in operation, now moved out of Wichterle's kitchen and officially adopted by the institute as its own. Wichterle, in turn, was invited to Rochester, where he gave briefings to different corporate groups about every part of the technology, ranging from mathematics to biology to commercial production.

Later that year, NPDC sublicensed to B&L the rights to the "gels," as the polymer lenses—actually one large, water-absorbing molecule—were called and to the manufacturing, which would involve two methods. One was centrifugal spin casting of the lenses in the wet condition developed on the Christmas apparatus in 1961. The other was a more precise lathe cutting technique. In this technology, the lenses are cut like thin slices of salami off a dry cylinder, put into water, and "hydrated." The dry slivers become soft and enlarged. Each dry slug can be of a specific size and prescription. With the rights to the Wichterle gels and manufacturing techniques, Bausch and Lomb further refined the production technology and launched a new business.

Greed

The price of success in the marketplace is everywhere the same: competition and thievery. By 1978, Bausch and Lomb's success had the greed glands of its competitors working overtime.

Pollak traveled to Prague to tell the Academy it must be prepared to honor the clause in the license agreement with NPDC to share the costs of protecting Wichterle's patents fifty-fifty. As in any U.S. corporation, in Czechoslovakia rights to researchers' work are assigned to employers. Wichterle's was the Academy of Sciences.

The patents under attack were U.S. patents assigned to the Czechoslovak Academy of Sciences, which had succeeded in obtaining U.S. protection for the Wichterle gels to prevent its proprietary work from being exploited in the U.S. market without payment. A patent is nothing more than a time-limited legal monopoly intended to reward those who have expended time and money to be innovative. As some inventions are relatively easy to reproduce, patents discourage competitors who might otherwise copy the technology.

Under the Paris Convention on International Copyrights, Trademarks and Patents, the Author's Certificates, issued to Eastern European and Soviet researchers, although carrying no monopoly rights in their own countries, are recognized internationally as equivalent to Western patents in terms of their technical novelty.

Continuous Curve, a potential competitor and subsidiary of Revlon, was about to launch a nullity suit against NPDC and codefendant Bausch and Lomb. The Academy's Patent Office ran some numbers and concluded that their share, if they lost, could be very large. Millions of dollars was at stake. Wichterle naturally believed that his patents were good and would withstand the challenge, but he wasn't in a position to guarantee what a U.S. court might decide. The Czech lawyers for the Academy were in a dilemma, since refusal to share the costs of defending their own patents could lead to nullification of the license agreement with NPDC, which stipulated joint defense.

NPDC agreed to renegotiate. In return for the Czechs' not participating financially in the suit, NPDC would be allowed to buy outright the patents under which it had been granted a license. The Academy was, therefore, ready to make two agreements: one voiding the existing license granted to NPDC for both the gels and the processing patents, the other a buyout contract. This meant that NPDC would become the owner of the patents, and the Czechs would give up, for all time, any proprietary claims. This second agreement constituted, in Wichterle's view, yet another contract entitling him to further commissions from the Academy. The Academy took the position that it was really a continuation of the same earlier agreement which had already made Wichterle the richest man in Czechoslovakia: no more commissions owed.

Wichterle responded by hiring a local lawyer and suing his employer, the Academy of Sciences. Wichterle won his case after a year, his victory based, in large measure, on the fact that the signatories were different. Though the individuals were the same, the buyout was with a new entity created by NPDC called NPD Technology. The original license agreement was with National Patent Development Corporation (NPDC).

After fighting for further commissions at home, Wichterle turned to preparing himself for the Continuous Curve attack on his wallet. It is not enough to have a good product to be competitive: a company must be able to make it as cheaply as possible. Continuous Curve was challenging Wichterle's patents over the issue of lathe cutting, the preferred method of manufacturing, which eliminated the need to smooth the edges of the lens.

Continuous Curve's nullity suit was a legalized fishing expedition to invalidate Wichterle's patents by any means possible. A patent can be declared invalid if it can be shown that the inventor has killed someone, that the invention is obvious, or that the idea has been stolen from someone else. Continuous Curve tried to use every legal argument in the book to attack the patent, including one that ultimately discredited their whole case. Wichterle spent weeks on the stand in the Los Angeles Federal District Court answering repeated questions about his morals, his police record, and his relationship with his coworker and coauthor Dr. Drahoslav Lim, the implication being that the lens was really Lim's, not Wichterle's, invention.

The most threatening attack against Wichterle's patent deserves a glance backward to its origins. During a train ride from Olomouc to Prague in the fall of 1951, Wichterle peered over the shoulder of his seatmate and noted an advertisement for a metal prosthetic eyeball in the ophthalmological journal he was quietly reading. Moved to declare what a terrible idea it was to stick such material into people, Wichterle piqued the curiosity of his seatmate, who had never thought much about the undesirability of sticking a ball of tantalum into an empty eye socket. During the next three hours to Prague, Wichterle and his new friend Dr. Pur had an energetic discussion about alternative prosthetic materials. Wichterle thought plastics had much more to offer and could be constructed to be totally compatible with biological tissue. The ensuing back and forth with Dr. Pur, who turned out to be the secretary of a high-level government health commission, led Wichterle to think out the basic requirements for a good biocompatible material.

These were, and still are, compatibility with surrounding tissue, permeability, absence of irritating impurities, non-similarity with natural proteins (to prevent immune reactions), and stability. These requirements hold for soft lenses, as well as any prosthetic implant designed for a long sojourn in the body.

Dr. Pur was so excited by Wichterle's ideas and his confidence that such materials were around if one looked, that a few days later, Wichterle received an invitation. The good Dr. Pur wanted his new friend to regurgitate the train conversation to the commission on which he served. To this particular meeting came leading ophthalmologists and officials of the Ministry of Health.

Wichterle reviewed the ideas he had developed in conversation with Dr. Pur, describing the kinds of materials and properties best suited for ophthalmic, as well as general, implantation. When they asked whether any of these wonderful materials existed, the important professors and doctors were dumbfounded to learn that the answer was no. How could this frivolous organic chemist dare to waste their precious time talking about nonexistent materials?

Stung by the dismissive reaction, Wichterle asked his assistants the next day what they thought about the concept of developing biomaterials. Only one of his five assistants was enthusiastic, but this was enough to trigger the search for the right material, which had to be optically clear as well as meet the other requirements for use in the eye and other body parts.

The search stopped six months later when coworker Lim synthesized hydroxyethylmethacrylate (HEMA), a material that is optically clear and easy to synthesize. The difficult part was finding the right combination of purity and shape stability. Without a stabilizing additive, or *cross-linker* in polymer parlance, HEMA would be as formless as cud.

A patent can be declared invalid if shown to be preceded by an earlier invention, usage, or information already in the public domain. A publication by Wichterle and coworkers that predated the lathe cutting patent was produced by Continuous Curve. The article described a method for cutting ultrathin sections of tissue samples embedded in non-cross-linked pure HEMA polymer. However, HEMA, the polymer material Wichterle's group had selected back in 1953 as the ideal one for use in the eye, was cross-linked to give it the necessary shape stability and toughness, the very qualities which had impressed Pollak. The arguments ultimately revolved around the definition of "new." Yes, a HEMA material had been described in the earlier article and cut in a dry form with a lathe. But without crucial cross-linking additives, it was simply not a substance which could be inserted into the eye. It was a different material—as different as chewing gum is from an automobile tire. Therefore, the said article did not represent a prior disclosure.

At this stage, NPDC's codefendant, Bausch and Lomb, got cold feet. Under the law, a defendant in a patent suit can change sides if he believes the case is a bad one. It is not considered the acme of ethical behavior, but it is legal and permits speedier settlements. As B&L was paying NPDC considerable sums of money in royalties for the right to use the Czech patents controlled by NPDC, should Continuous Curve win the case and B&L be on their side, then B&L would be able to negate its own license agreement with NPDC and save itself millions of dollars in royalties. Smelling defeat in the air, B&L decided to switch rather than fight.

A final attack on Wichterle's and NPDC's position was the

attempt of Continuous Curve to knock down the patent on the basis of the legal argument of nonpracticability. The plaintiff had dug up one Dr. Maxmillian Dreifus in Switzerland. Dreifus had been employed as an assistant to one of Czechoslovakia's leading ophthalmologists at the Second Eye Clinic in Prague, where the initial tests on animals and, later, human patients were conducted.

After 1968, the ambitious Dreifus and his wife decided to seek greener pastures in the West. They emigrated to Zurich, where he set up a private practice. From Dreifus, Continuous Curve's lawyers learned something interesting. It was common practice to destroy all medical records after ten years. The work which Dreifus had done for Wichterle was more than ten years old. As the main clinical investigator, Dreifus was in an excellent position to testify to the practicability of Wichterle's lenses.

In May 1977, Continuous Curve produced its surprise witness, who testified that the lenses he had tested for Wichterle were only suitable for animals. The things only looked like lenses from a distance, but were not designed for human wear. They had only been tried on rabbits.

With this testimony, the court was confronted with directly conflicting stories. Judge Laughlin E. Waters ordered that none of the witnesses could leave Los Angeles for three months, during which time the court would determine who was lying. Wichterle's attorneys knew that the only way for them to disprove Dreifus's testimony was to obtain the original clinical records, and this could only be done if Wichterle could go back to Czechoslovakia. By persuading the Czech Embassy to cooperate in not permitting an extension of Wichterle's one-month visa, the defense induced the court to make an exception and permit him to return to Prague.

In Prague, Wichterle got in touch with Dr. Černy, an organic chemist whom he had hired to work on the gels after 1968. Černy had been fired by the Technical University for his political activity in 1958, during the Stalinist period, and could not find a job. Wichterle eventually employed him as a laboratory assistant to help improve the method of casting lenses. In this capacity, Černy worked closely with Dreifus, supplying him with new lenses to try out on patients.

When Černy heard about Dreifus's testimony from Wichterle, he took a leave of absence from his job to help locate the crucial records. Černy learned that the records, if they still existed, were likely to be found stored in the monastery of St. Ignatius Church in Prague. Two weeks of desperate hunting in dusty piles of discarded clinical records produced eighteen patient cards written in the hand of Dreifus.

The discovery was treated as top secret: Dreifus and his lawyers were to be ambushed. When the court reconvened, the defense led Dreifus through his earlier testimony one more time. Yes, the lenses were only suitable for rabbits, unfit for human eyes. These things were not real contact lenses, and so forth, Dreifus repeated for the court.

The defense showed Dreifus a portion of one of the patient cards which did not have his handwriting on it. Yes, that is a patient card from the Second Eye Clinic, testified Dreifus. Then another portion of the card with only his handwriting on it was shown. Yes, that was his handwriting. The trap was sprung. With these admissions, the defense then read the patient information Dreifus had written on the cards. When asked whether the notation "She showed no irritation and took the lenses home" referred to a particularly dextrous rabbit, Dreifus and his lawyers fell into a state of consternation.

Seeing the collapse of their witness under the piecemeal revelation of the defense, the plaintiff's lawyers called an emergency recess, charging that the case was creating physical danger for the Czech witnesses. Lives were at stake because of the political ramifications back home.

Although that was a transparent fabrication, a recess which gave time for Continuous Curve's lawyers to cook up a new line for the flustered Dreifus was granted. When Dreifus returned to the stand the next day, he maintained that he had deliberately falsified the reports in order not to disappoint his friend, Wichterle, to whom he knew the lenses meant so much. All he wanted to do was make the nice scientist happy.

The case ended quickly with NPDC's getting damages and Wichterle's integrity vindicated. On the desk of Pollak and Leidig today is a plaque inscribed with the words of Judge Waters's summation: "THIS COURT HAS NEVER IN ITS HISTORY SEEN SUCH A DISCREDITED WITNESS." The defense considered pursuing Dreifus for giving false testimony, but Wichterle was not hungry for revenge. If anything, Wichterle felt Dreifus should have been rewarded, for his duplicity won the case for their side, with an award of triple damages.

Visit to Strazisko

Although I had picked up bits and pieces of the soft lens saga during meetings with Wichterle in Prague, to learn more about this unusual man, I followed up on an invitation to spend a weekend with his family in July 1986. The stay also afforded a unique opportunity to gain insight into a man who speaks with unabashed candor and honesty, lives comfortably and prosperously, and is a "privileged" person still in many ways, despite past political transgressions.

Moravia is a part of contemporary Czechoslovakia whose name in the seventeenth and eighteenth centuries was associated with a movement of worldwide influence, a form of "soft" technology transfer. The Moravian Brethren, under the leadership of Jan Comenius, carried yet another form of Protestant piety, pacifism, and learning to the shores of the New World. They settled in Bethlehem, Pennsylvania, and Winston-Salem, North Carolina, and eventually extended their congregations across much of the United States. Comenius was a man of great learning whose reputation as an educational reformer was known throughout Europe. His innovative ideas included belief in the use of the vernacular, instead of Latin, in the classroom and free universal education for both sexes, as well as for the deaf, dumb, and blind. There is disputed evidence that John Winthrop, Jr., offered Comenius the chance to succeed Henry Dunsten as president of Harvard College.

Today, Moravia is known for its wines. It is also the birthplace of such figures as Gregor Mendel, the Augustinian monk who pioneered the science of genetics, and Kurt Gödel, who disturbed the certitudes of the mathematical world by establishing its inherent indeterminateness. It has two important industrial centers, Brno in the south, and Gottwaldov in the north. Before World War II, Gottwaldov, formerly Žlin, was the hub of the largest shoe-making empire in Europe: Bata. Between these two industrial poles is gentle rolling farmland. The villages of Moravia seem a little brighter and more prosperous than other parts of Czechoslovakia. Flowers provide splashes of color; houses, many with garages, are privately owned. Czech country villages are down-in-the-heel versions of their Austrian counterparts.

The relaxed atmosphere of the Wichterles' summer home is immediately set by the Ping-Pong table that greets the visitor crossing the threshhold. Tennis rackets and old clothes hang in a hallway which is actually a two-story atrium in the middle of the house. To the right of the entrance, a stairway winds to the second-floor interior balcony, off which three large bedrooms face south across the valley. Seven Wichterle grandchildren and frequent visitors prevent the dust from settling.

After showing me my bedroom, a Pullman-sized alcove off the first landing, Wichterle took me for a stroll through their six-acre section of pine-forested slope, which he lovingly refers to as his "garden." Not noticeable from the street below is the string of footpaths which have been set in a meandering series of interconnected switchbacks leading to the top of the hill. Along these strolling paths are wooden benches and half-logs from which to look out into the valley below. In one remote offshoot of the path, a showerhead rises pensively from the ground, next to it a log and bar of soap. This was to become my bathroom of choice.

The hillside is managed and maintained by Wichterle & Co. Thinning and cutting are all done by hand—no riding mowers or weed whippers. These suburban frivolities are hard to come by in socialist countries, where sickle and scythe are still used to cut grass. The garden is the family labor of love. The pine trees are under severe ant attack. The Wichterles have to prune out dying pines constantly, providing their ceramic stove with a regular source of fuel. From the government's standpoint, the trees are worth more than the property. There are no taxes on the value of the land, only a tax on the value of the trees. In Czechoslovakia, there is no legal limit on the number of houses anyone can own, although there is a restriction that individuals cannot own houses with more than five bedrooms.

In a busier village, the Wichterles' home would be engulfed in traffic and noise. Established in the 1860s as a weekend retreat for the well-heeled burghers of nearby Prostejov, Strazisko hasn't changed much since 1925, when Wichterle's father, Charles, purchased their summer home. A new all-weather tennis court across the street from the Wichterles' house is being built as part of the community recreation center. The fish in the stream across the street where young Otto used to hang a pole in the summer do not jump anymore, but much else is still the same.

A quarter of a mile across the valley is the same house his wife, Linda's, family used to live in in the summer. Next to it, set in an apple orchard, is the clay tennis court where Wichterle played with her parents while she chased balls. Still standing one hundred yards from the tennis court is the communal two-story summer dormitory built and owned by seven families from Prostejov. This mini-apartment complex, a kind of nineteenth-century condominium made from hand-hewn wood, is still occupied every summer by the descendants of the same families of one hundred years ago. Little has changed. Water is carried in buckets from a nearby hand-pumped well to supply all their cooking and washing needs. Only electricity has been added.

As we strolled back down the garden path, Wichterle reflected that the people in Strazisko have a pretty good life. The air is clean; there is little stress. During the war, people here were barely affected by the Nazi occupation, and today, they are little affected by the new system. Life goes on pretty much as before.

"It is very hard to change anything," Wichterle observed. "People here are strong and independent-minded, but life is easy. No taxes to pay, no mortgages to worry about or need to scrimp and save to send children to college. The problem with Czechoslovakia today is that people have it too easy. Life is too comfortable. Workers have complete security, and they do not have to work very hard. Administrative work is a joke. In the old days at Bata, there used to be glass partitions between work areas. Toilets had half-doors. No one could goof off easily. Today, a Czech office worker loves his privacy more than anything else."

How about scientists? I interrupted. Are they better off than before?

> After World War I, under the First Republic of Tomas Masaryk, the Czech government didn't support science to speak of. There was an Academy of Sciences, but that was like the British Royal Academy—more of an unpaid debating club. There was little chemical industry. Except for sugar production, some dyes, and additives for the leather industry, there were no industries which required much chemical science. Only after the Communists took over did science get strong government support. By adopting the Russian model, the Party made a commitment to support science at the highest levels.

Over our risotto dinner, I asked Wichterle about his impressions of America during twenty years of periodic visits.

"America is like a totalitarian society," he offered as an opener. Asked to explain, he gave a Menckenesque description of intellectual and social conformity, political blandness, and herd thinking. This, at any rate, was his assessment of suburban America, where he had had the apparent misfortune to spend most of his time when not delivering speeches or consulting for U.S. firms.

"There is less diversity of thought in America than Russia, even if it is self-imposed in America. In Russia, after people get comfortable with you and take you into their homes, everyone in the family has an independent opinion. I didn't have one good political discussion in America." Wichterle attributed this blandness in part to the lack of a tradition of real political opposition. The great uniformity of ideology which many would regard as America's strength was for Wichterle stultifying. The standardized housing of suburbia symbolized in his mind standardized thought. "People in the suburbs live by very strict rules, even if they are unwritten. Women have to buy a new dress every two weeks. In the old days, only a parvenu did this. Women of class had only one good outfit, but this was perfect and it was worn for all important occasions."

It was therefore no surprise that cocktail parties were among his pet peeves, along with apple pie. Corn on the cob, however, made a favorable impression, so favorable that he took home with him a newly acquired taste difficult to satisfy in a country where corn is consumed only as hog feed.

One day, Wichterle was seized by a craving for some good old American-style corn on the cob, as he drove past an open field of Moravian maize waving in the breeze like Old Glory. Not a man to be cavalier about communal property rights, he drove up to the office of the local cooperative and inquired whether he might take some corn. Asked by the manager how many wagonloads he wanted, Wichterle indicated that he only wanted to take a few ears. What could he possibly want with a few ears? When told that he wanted to eat the corn, a dumbstruck look spread over the manager's face. As Wichterle related the story, the manager acted as if he'd been confronted with a freak of nature, calling in his buddies in the next room to take a look at this strange creature who actually wanted to eat corn. Slack-jawed, but reassured that he did seem otherwise normal, the manager told Wichterle to go help himself. Wichterle insisted on paying. The manager insisted not. What if someone sees me pick the corn? countered Wichterle. They will think I'm stealing cooperative property. Just tell them you are from the state inspector's agency taking samples. His sense of propriety satisfied, Wichterle passed quietly into annals of local agricultural folklore.

Strong Genes

Like his American contemporary John Jacob Astor, Francis Wichterle got his start selling animal hides, widely regarded as a Jewish business. By trading fur skins, Otto Wichterle's grandfather accumulated enough capital to start his own farm implement business, which prospered well enough to attract competition. A Viennese Jew named Zaiček decided to establish a rival farm equipment business in Prostejov. This occurred at a time of growing tensions between German nationalists and Czechs desiring to have their own language and culture granted equal status to the German.

These ethnic stresses served Francis Wichterle's ends. By using a blatant "buy Czech" competitive campaign, Wichterle, a Czech nationalist with a German name, drove out of business Zaiček, the German Jew with a Czech name. Such were the ironies of the Hapsburg melting pot.

By World War II, Wichterle and Kovarik, as the firm was then called, had grown to four thousand employees but was also 70 percent owned by the banks. After nationalization, Wichterle's brother continued to manage the business, but instead of working for the banks, he was working for the Ministry of Machine Building.

Otto Wichterle shares with two sisters an ascetic, uncompromising determination and independence of character. A sculptor and amateur Sanskrit scholar, Hana Wichterlova refused to take her cues from the Party's artistic authorities during the Stalinist era of the 1950s and early 1960s. "Good taste" then demanded socialist realism, a form of art similar to what one sees in the AFL-CIO Washington, D.C., headquarters. Hana persisted in her nonconformist, subjective style of sculpting with the result that her work was banned from public display. Like her brother, Otto, Hana was sympathetic to the Communists before the war, but she, too, possessed an independent spirit. Rather than change her style of art, Hana chose near-starvation, refusing to take any financial help, even from her family.

Sticking to her principles ultimately paid dividends. When attitudes about artistic experimentation started to change in the early 1960s, pre-Dubček, Hana's work was in great demand. She became widely acclaimed and, by Czech standards, wealthy. Today her work can be seen at the Zbraslav National Gallery outside Prague.

Emy, his second sister, was strongly influenced by their father, Charles, a devout Catholic and loyal supporter of the monarchy. Unlike her agnostic siblings, Emily Wichterle accepted her father's faith. She joined the poorest of the prewar Catholic orders, the School Sisters of St. Francis, which continued in existence until 1968. After 1968, the order was disbanded by the government but continued to function in the capacity of a charitable organization. Living only by the charity of others, the order continues to serve the old and needy.

A man of strong convictions, Otto Wichterle never played professional politics. The pursuit of power is not on his agenda, nor is he now ideological in his approach to life. For Wichterle, the goal of politics should be to strive for "socialist decency." *Decency* is a favorite word of Wichterle's. Decency and the elimination of small-minded pettiness would characterize the world that Wichterle would have. His disenchantment with Communists began after the war when he saw that many acted like Nazis in their treatment of opponents. Wichterle's leftist activism as a student in the 1930s was enough to qualify him for an extended period of Gestapo interrogation in 1942. Unable to find any proof that he had actually been a member of the Communist party, and intimating that he could be useful to the German war effort at the Bata Research Laboratories, where he was working, the Gestapo released Wichterle after he had been imprisoned for four months. This was not to be the only time Wichterle's scientific value to society would protect him.

His desire to remain independent of professional politics extended to refusing even nominal payment from the government for his participation in parliamentary deliberations during the Dubček period from 1968 to 1969. He and other scientists were asked to contribute their expertise to the formulation of scientific and educational policy as ad hoc parliamentarians.

Two months prior to August 1968, Wichterle and four others had expressed their political convictions in a document which came to be known as the "2,000 Word Manifesto." Written in June, during the period of openness and self-criticism, this document attacked the Party for abuse of power. It was published in four daily newspapers, including the Party newspaper *Rude Pravo*. Party conservatives branded the tract counterrevolutionary. In the spirit of the "new openness," Wichterle and his coauthors defended their position on TV.

After the Soviet invasion of August 1968 brought the "Prague Spring" to a close, Wichterle continued to serve in the Parliament as a member but added to his reputation as *enfant terrible* by openly protesting what had occurred as violating various human rights charters. On April 1969, Parliament was supposed to be dissolved and new elections called. Instead, it voted to extend itself for one year. This was absolutely illegal, according to constitutional principles.

Wichterle and Kriegel, a Central Committee member close to Dubček, demonstratively voted against this by standing on their chairs in Prague Castle, seat of the Parliament. With this final act of principled protest, Wichterle resigned from the Parliament. A couple of months later, he was informed that the Academy was going to elect another director of the Institute for Macromolecular Chemistry.

According to procedure, the Academy ran a notice asking for applicants to fill Wichterle's vacancy. No one applied except Wichterle. For three years, in an impressive show of solidarity for Wichterle, not a single application was filed for the job. Only in 1972 did the Academy finally appoint a new director.

Wichterle is philosophical in his recollections of these times. There were advantages, for him and for the institute. A growing concern he had after eight years of directing the institute was attracting new blood. The year 1968 produced vacancies in the research staff which had to be filled. For the most part, the best people didn't leave. Relieved of his administrative duties, he had more time to indulge his great love—tennis. Yet he also held the position of senior scientist, continued his research unabated, and, after a period of time, resumed his travels to the United States.

Even after his dismissal, the government never tampered with his legitimate financial rewards stemming from the licensing of his soft lens technology. In this regard, Wichterle thought the Party behaved "decently."

At a trim seventy-six, Otto Wichterle leads a life of active retirement. He has recently had patents issued for highly water-absorbent lenses and a device for accurately measuring eyeball size for fitting lenses. A Czech movie is to be made about Wichterle, Hollywood-fashion, which will overhaul his personal life by suppressing the tedious fact of a happy thirty-eight-year marriage to his childhood sweetheart. In the pre-AIDS society of Czechoslovakia, the new twin Western fashions of family and fidelity have not arrived yet.

Soft lenses are now manufactured worldwide, thanks to the sparking of ideas that took place on the three-hour train ride from Olomouc to Prague in 1951. Few Czechoslovaks, however, wear soft lenses. Only thirty thousand fittings per year are performed for domestic consumption, over one million going to the USSR. Czech optometrists simply are not interested in offering soft lenses to customers. The pricing system has not recognized the value of the tedious and time-consuming service of fitting, refitting, adjusting, and readjusting lenses for each sensitive eyeball. "We are suffering from a kind of technical encephalitis—our brains are overdeveloped in relation to the capacity of our economic systems to execute," commented a scientific friend of Wichterle's, Gyula Hardy, in neighboring Hungary. He could have been talking about all the Communist systems.

Yet there are islands in the vast swamp of economic lethargy that wears down the person of merely average energy and ambition. If Wichterle was an individual hummock who inspired and energized others, the United Agricultural Cooperative, Slusovice, is an island of continental proportions.

Slusovice

"Slusovice is Bata." This is not officially recognized, of course, nor does Frantisek Cuba like the comparison, as it only makes life more difficult for him. Yet for the father of Joe Maly, my Polytechna associate with whom we spent several hours sipping wine on the way back from Slusovice, where the giant agricultural cooperative is headquartered, there was not a hesitation of doubt when I asked him about the nationally renowned business empire.

The Bata name is a haunting ghost to Czechs at the highest level. It is a source of great historical pride and was a model of industrial accomplishment during its prewar era, being to the

European shoe industry then what IBM is to the computer industry today. After a false start making slippers in 1904, Tomas Bata emigrated to the United States, where he worked in several shoe factories and studied American methods. He returned to Moravia in time to take advantage of the need for providing the Hapsburg army with footgear for World War I. In 1917, he was producing ten thousand pairs per day, and by 1939, Bata employed sixteen thousand workers in Žlin, producing hundreds of thousands of shoes daily.

Tomas Bata combined strict production discipline, a high degree of mechanization, and hard work with all-embracing company paternalism. Žlin, renamed Gottwaldov after World War II in honor of Czechoslovakia's first Communist president, Klement Gottwald, was turned into a company town. Bata's paternalism took the form of a host of facilities and services which he provided his employees and the town of Žlin. He built over three thousand low-rent houses for his workers and established training schools, hotels, restaurants, and hospitals for the community. "We want to make better human beings—healthier, better educated, richer and more self-confident" was a stated Tomas Bata management goal.

Bata used the safest and newest production equipment, believing that a humane working environment was essential for increasing worker productivity. In exchange for his interest in employees' welfare, they were expected to adopt a collaborative attitude with management and regard themselves as "partners." Expectations were high and discipline strict. Each workshop had precise "daily plans" and was expected to maintain strict time budgets for each task.

Profit sharing with workers was part of the Bata philosophy too. If production within a specific workshop exceeded the minimum level, employees of that workshop retained one-half of the profit, the rest being reinvested in Bata. When Wichterle worked there during World War II, directors of the company made only four times the wage of the line worker. With profit sharing, workers often made as much in salary as directors. Bata believed in cooperation, not struggle between managers and workers and instilled a spirit of enterprise through training, exhortation, competition, and higher earnings linked to achievement.

In 1949, the United Agricultural Cooperative was founded in Slusovice, a suburb of Gottwaldov, still saturated with the Bata legacy. When Frantisek Cuba took it over from his father in 1963, the total sales was 1 million crowns, or approximately $200,000. In 1987, turnover of the United Agricultural Cooperative was 5 billion crowns.

The cooperative is spread over seventy-five thousand acres of northern Moravia, incorporating a wide range of agricultural and related operations. Slusovice has an active livestock breeding program and has developed expertise in artificial insemination techniques. It grows corn, makes its own specialized agricultural machinery, and produces its own microcomputers. They make their own fertilizers and pesticides. The cooperative processes and packages their own produce. The finest grocery store in Gottwaldov, owned by Slusovice, would rival any European counterpart in appearance and selection. Slusovice supplies consulting services to other cooperatives in Czechoslovakia and to agricultural enterprises in the Soviet Union and China. The United Agricultural Cooperative Slusovice provides all-embracing services to its cooperative members: recreational facilities, health care, housing, credit to build houses. Slusovice achieves in two years of growth what is expected of the average Czech industrial enterprise in twenty years.

These were the proud facts that were being rattled off by Frantisek Cuba, son of the founder and nationally known entrepreneurial hero, as we sat in the festooned alcove of his comfortable office. Blue ribbons, awards for outstanding achievement in livestock and corn production, were hung all about. Nor was his office ordinary. Upon entering the reception area, the first impression is one of Western efficiency. Touch tone phones and word processors abound. In his own office, the desk is oddly small in proportion to the man. Robustly built, Frantisek Cuba (pronounced "chewba") is in his early fifties and has a smooth, bespectacled baby face and an Elvis Presley pompadour. According to the mythology of Frantisek Cuba, he has never taken a vacation in twenty-five years.

After a half hour of discussion, it was clear that Cuba was growing impatient, and it was time to begin the tour. In marched a ramrod executive whose job was to show me the facilities that I had expressed an interest in seeing. These included the animal breeding operation, fertilizer production facilities, the computer factory, and the soil analysis laboratory.

As I left the office, I asked Cuba about his contacts with American companies. It seems there are few. They buy some chemicals from Stauffer and Monsanto. "Americans are not flexible," commented Cuba. "Twenty-eight percent of the world's trade is barter. Yet the American attitude is one of inflexibility. 'If you have money, we'll deliver; otherwise, don't talk to us.' This is the American attitude."

As to Bata, Cuba rejects most comparisons. "I studied his system, but ours is based on a different principle. His was based on a very strict regime. Workers had to clock in exactly, adhere to strict working hours, follow the line exactly. Our aim is to create freedom for people to make their own decisions and exercise initiative."

After my interview with Cuba, I was taken to the auditorium and shown a slick audiovisual presentation of Slusovice displaying its achievements: ten thousand embryo transplants annually; minimum weight gain of their cattle of one kilo per head per day; apologies for the pictures in which workers are shown manually repairing broken tractors (everything is now automatically welded by robots). Slusovice produced twelve hundred TNS microcomputers in 1985. These are used for data processing, dosing, greenhouse control systems, and regulation of laboratory fermentation equipment.

The cooperative manufactures their own soft drinks and packages their own food. The Slusovice racetrack is a source of entertainment for the workers and income for the cooperative.

It was clear that in Slusovice trains do not only run on time, but ahead of time. In every facility, we were punctually met by employees who snapped to attention and gave brisk explanations of their respective operations.

Over the course of the afternoon, Mr. Latoka, director of the Chemical Unit, provided some insights into Slusovice. Managers are evaluated every quarter on a 1 to 5 rating scale. Their performance is reviewed principally on the basis of three considerations: quarterly profit, technical innovativeness, and ability to solve people problems. If the manager's rating falls below 2.5, he or she is demoted. Often, the problem is one of matching the manager to the right job. Demotion usually means transfer to a different unit where one may rise again to a position of former responsibility. There are second and even third chances in Slusovice. Over the fifteen years Latoka had been there, he was aware of only three cases in which managers had been fired summarily for poor performance.

New employees are hired out of the universities on a six-month trial basis consisting of two phases. During the first three months, everybody shovels dung or performs some equivalent manual labor. During the second three months, the candidate works at different places requiring different technical skills. After six months, there is a psychological test. The basic qualities that are being looked for are attitude toward work, initiative, and technical knowledge. The results of the six months screening process are a rejection rate of 40 percent and an acceptance of 40 percent. Of the remaining 20 percent, half get jobs different from those for which they applied, and half get jobs at a lower level than they applied for. Of the 40 percent who are accepted initially, 60 percent quit after six months. For most, the pace is simply too fast. Senior people transferring in, like Latoka, are not subjected to this process but are rigorously evaluated. Managers at the same level may have different salaries because of different management ratings. In general, Latoka believed that the higher up in the organization, the less important was pay and more important the rewards of doing a good job.

As we crossed a concrete bridge to the embryo transplant laboratory, Latoka pointed out that this had been built in sixteen days, working night and day. Normally, this kind of construction project would take nine months to complete. Around the environs of Slusovice it is not uncommon to see a sign that says "LET CUBA BUILD IT." This, says Latoka, is one of their important rewards: the townspeople's appreciation of Slusovice's efficiency.

There are eighteen business units, each an independent profit center. If a unit is failing, it either goes out of business or gets credit to keep operating. One to two business units per year fail or are reorganized. So many visitors go to Slusovice that the co-op is creating a travel and tourism business unit. Each unit has control over its profits, except for a portion that goes to the general cooperative coffers. The retained earnings can be spent at the discretion of the unit—for higher salaries, housing, new equipment, or whatever the unit management considers most important. Director Cuba has no power to interfere in these decisions.

There are three grades of employees in Slusovice. There are those who do not have to come to work at all if they can get the job done from bed or while fishing, those who have a flexible schedule, and those who have a fixed schedule. No drinking is allowed until after 8:00 P.M. in Slusovice. If Cuba wants to see somebody on short notice or have something done, he does not want to have to deal with a boozy worker.

Cooperative equipment can be used by members, especially for doing plumbing and electrical work at home. Enlightened self-interest is behind this policy. The faster the workers can get their plumbing installed and electricals wired, the easier it will be to get their minds focused on their work. These are services not readily available in Czechoslovakia and subject to the "Harry Homeowner" approach. In fact, everybody in Czechoslovakia is "Harry Homeowner." Do-it-yourself kits would be a great business in socialist countries.

Concentrating on the job is the key ingredient in the success of Slusovice. Working fast and intensively is the norm. To make this easier, the cooperative rewards people so they can concentrate on their job and not be diverted by more mundane needs. Once you make the Slusovice team, you have the benefit of the doubt when problems arise. There is no sick leave per se. If there are personal problems, workers take time off with no questions asked, according to Latoka.

The soil analysis laboratory was well-lighted, Disneyland-immaculate, with green plants abundantly evident and plenty of top-of-the-line Zeiss Jena analytical instruments from East Germany. Slusovice has developed a highly scientific system of soil sampling and analysis, providing clients as far off as China and Lithuania with recommended programs for soil treatment and livestock breeding. This total systems know-how is packaged and offered on a consulting basis to both foreign and domestic customers. Latoka said that this business came from the Western journalists who have visited Slusovice. *The Financial Times* and *The Washington Post* with their international readership have been particularly effective promoters of Slusovice expertise.

Over a late lunch at the co-op hotel and restaurant complex, Latoka explained that he joined the co-op fifteen years ago after being a frustrated R&D chemist at the Pardubice chemical plant. He had many good ideas which were simply stuck in his drawer gathering dust. Someone at Slusovice, however, had read about his work with micronutrients and thought it could be interesting. After receiving a visit, Latoka accepted a job, and within one year, Slusovice was making money with his ideas of applying small doses of various soil nutrients to improve yields.

The extent of the efficiency consciousness of Slusovice was illustrated over lunch by an example of the way polyvinyl chloride (PVC) bags are processed. Plastic fertilizer bags, after being emptied of their contents, are washed, hosed down in a manner so that the residual fertilizer dust is sluiced off into the fields to provide a low-intensity fertilization bath. The clean plastic bags are then recycled and granulated. This granulate is then traded to the Austrians for equipment used in a joint venture to provide equipment and expertise for a corn growing project in the Ukraine. What the Austrians do not take is used to make food packaging containers for the cooperative's food products which are sold through their own stores.

Slusovice has its own hard-currency account and can conduct foreign trade on its own, as if it were a state foreign trade company. Managers at Slusovice travel all over the world to learn, to buy and sell, and to provide services. Latoka, who had traveled to sixteen countries, including the United States, was most impressed by the efficiency of the Danes, least by the Brazilians.

Slusovice combines vertical integration and self-sufficiency with aggressive expansion into new areas of profitability. The speed with which they complete construction is legendary and has given rise to a construction unit. Slusovice built the largest department store complex in Prague. The co-op is building advanced rotorless fermentors developed by the Czech Academy of Sciences for both domestic use and export. Many of the activities of Slusovice, such as machinery building, construction, fertilizer manufacture, and microelectronics, are driven by the desire to be independent of the supply system of the economy as a whole. Dependence on "the system" will only slow Slusovice and dilute the quality and efficiency of its operation. Whenever there is a business opportunity that will further reduce dependence on the standard supply system, Slusovice is poised for action.

What is the source of Slusovice's phenomenal success?

"Cuba surrounds himself with good people. He's not afraid to have around him those who are smarter and wiser than he is." "Cuba is well connected in the highest circles, but unlike others with good personal connections, he uses his for the good of the co-op, not his personal benefit."

Such are some views of Cuba's success which go beyond energy, drive, and intelligence. Cuba is frequently on Czech television. In response to a similar question, as to how he managed to "get away" with all his success, he attributed it to good lawyers. Slusovice is subject to controls and oversight by thirty-one different agencies and ministries. The cooperative is constantly monitored. His lawyers are told to go to the extreme edges of the law, but not beyond. Like Gorbachev today, Cuba made his reputation by stretching the system and probing its ability to accept new ways of business.

Technology Transfer

Music

Technology transfer is another name for communicating knowledge and technique to accomplish a job. Whether cooking, writing, management, or horsemanship, each can be the subject of an approach or method. The instructional and everyday nature of technology transfer, as well as its cultural dimension, was illustrated by the efforts of a young Czech musician to adapt to, and teach in, a new environment.

On the evening I returned to Prague from Slusovice, I had dinner with Slavka to hear about her trip to the United States. She was a piano teacher I had gotten to know through a mutual friend. When I first met her, she had only one dream, and that was to visit America. The dream came true when a Czech-American tourist she had met in Prague invited her to Taylor, Texas, for six weeks. There was a small community college at which to teach piano lessons, providing an official reason for the trip.

I'm glad I went, but I'm glad to be home. The experience was totally different from anything I'd known in Europe. I was shocked. It's very hard for me to explain to people the feeling of alienation, the differences. I realize, of course, that Taylor, Texas, is a small town and is not all of America. It is not typical. But first I was disoriented by the physical differences. The town had no center. There was no organization, just a sprawl. All European towns have a square, a church, a town hall—a center. There weren't even any sidewalks!

The people I stayed with were very nice, but all they did was work. Luba and her husband worked twelve hours a day. She had a travel agency, and he was a doctor, so I was alone a lot. It was very difficult for me to meet anyone. There was no sense of community, no history. People don't sit around and talk at restaurants. It's eat and run. Time is money. Everybody is only interested in work and business. It seems Americans work all the time—I guess that's why you're rich. Of course, no one was the slightest bit interested in what I did, so that didn't help.

Trying to teach Americans to play the piano was an interesting experience. I was surprised at the students' complete ignorance of music history or culture. All they want to do is play music. I tried to explain that the quality of the sound depends on the way their fingers hit the keys, but they didn't care about quality, just what was easiest. In fact, everything is made so easy, people don't have to use their heads. Instructions are written on everything like all people were small children—turn left to open. I was really surprised.

I was so unhappy where I was, Luba arranged for me to work in a rib restaurant in Eagle Springs, Arkansas, where she had a cousin. That was the best part of the visit, because I had a chance to meet people working in a restaurant. It was easy to get into conversations. The customers would notice that I had an accent and ask me where I was from. I had a lot of long conversations, because people always asked me how I liked America and whether I was going to go back home. I shocked people by saying I was going to go home happily and that I didn't really like America that much—at least what I'd seen of it. People didn't understand when I said all Americans do is talk business.

Americans, of course, are very friendly, very helpful, but not

> easy to make friends with. Their material well-being is fantastic. You can get anything you need, but I could never be happy living your way of life.

Transferring technology is a delicate business, much like transferring plants from one soil to another. Like horticultural transplants, technology transfer is highly dependent on matching conditions. If these conditions are not right, the technology will die, no matter how good it is. In Taylor, Texas, the local cultural preconditions for receiving Slavka's "European approach" to piano instruction were not favorable. With time, she might have prevailed.

Electroslag Welding

In transferring technical information, there must be a similar technical infrastructure within the recipient organization that can readily grasp and appreciate the knowledge that is being acquired. Beyond technical compatibility, the key ingredient for successful use of foreign technology is commitment—and that too can change as circumstances do. Key people die or leave. Market conditions change. The efforts of a small Cincinnati company to acquire Czechoslovak welding technology illustrate the problem.

I first learned of the Research Institute for Welding in Bratislava when a U.S. producer of steel mill products called to say that it had identified some interesting technology for resurfacing rolling mill rolls. The firm, which prefers to be called XYZ, had learned this as a result of searching on-line technical data bases of the American Society of Metals. However, they had no idea how to follow up or whether it were even possible to acquire the technology. Czechoslovakia, after all, was behind the Iron Curtain.

Curiously, in Czechoslovakia and in certain other Communist countries, technology is sold the American way: one-stop convenience shopping. To buy technology anywhere in Czechoslovakia, a person needs only to know one address and one telex number—those of Polytechna. It is actually an excellent system for the seekers of technology. Polytechna acts as the official door opener to any institution in the country, be it in industry or academia. Through Polytechna, all meetings can be arranged and coordinated. The problem with such a system is that if it does not work, there is nowhere else to go. For owners of technology who are eager to sell, Polytechna's monopoly position has fewer advantages.

Informed of our client's interest in the technology at VUZ (the acronym for Research Institute for Welding) for resurfacing steel rolling mill rolls, Polytechna arranged a trip for me to Bratislava in Slovakia. The VUZ is a large, drab four-story building that covers a whole block on February Street on the outskirts of town. Work starts at 6:00 A.M. At my first meeting, I was given a tour of the facilities and a report on the current status of the technology. The so-called electroslag technique could be used generically for building up a new surface around any shaftlike object.

The problem is not an easy one. It is analogous to retreading a tire, except that metal is being overlaid on metal and, like a tire, must bond with the base material to form as homogeneous a whole as possible. Success in this "retreading" means considerable material and cost saving, especially when old rolls can be resurfaced to last as long as new ones. Poor adherence of the new material will result in failure under the extreme stress to which rolls are subjected in a mill.

The technology for achieving good metallurgical bonding to the base roll material lies partly in equipment, but mostly in good metallurgical modeling and understanding of stresses and solidification rates. The recipe is as important as the utensils. The recipe, in this case, was a series of computer models which fortunately were run on IBM-compatible machines and could easily be used by XYZ. The Eastern bloc, like the rest of the world, follows Big Blue.

The VUZ researchers were delighted to have the interest of an American company, especially for the roll application, which they, too, had been promoting. It validated their own belief in the merits of the technology whose introduction to Czech steel mills was frustrated because the mills lacked the economic incentive. Even though resurfacing a roll which would otherwise be scrapped would be much cheaper than buying a new forged roll, the steel mills could treat new rolls as a capital expenditure, money which was allocated by the ministry and viewed essentially as grant money. Hence, "new roll" costs were not an operating expense, but paid for out of budget grants from the ministry.

After I briefed our client, the company decided to send its own people to the institute for face-to-face discussions. During this first trip in 1983, XYZ learned that a man from Teledyne's Ohio Steel Division had visited the facility several years earlier and had an opportunity to study the technology in considerable detail, including taking snapshots of the equipment for rebuilding rolling mill rolls. Teledyne's man never returned and never paid Polytechna or VUZ a dime. Several years later, a U.S. patent was issued to Teledyne for an electroslag welding system closely related to the one VUZ had developed.

When it was clear that XYZ meant business and was willing to spend money for VUZ to make rolls an initially skeptical attitude dissolved in favor of enthusiastic cooperation. A test agreement was implemented in which the VUZ would overlay rolls according to the firm's specifications and send them to Cincinnati for evaluation in a customer's mill. In parallel was an option agreement which gave XYZ exclusive worldwide rights for two years to evaluate the technology, after which they could exercise a license.

Immediately problems arose. VUZ could not procure within any reasonable period of time a crystallizer, a kind of copper basin which surrounds the roll at the bottom. At the top of the vertically suspended roll is a "head," or crown, onto which are suspended long metal electrodes. These electrodes serve the same purpose, but are three inches in diameter, are twenty feet long, and weigh over a thousand pounds. They conduct current and provide the metal "fill." Rather than joining pipe or fixing a broken axle, the metal is melted into a chemical "slag bath" contained in the copper crystallizer. The electrodes must be of the same metallurgical composition as the base material. The liquid slag acts as a contact for the electrode, which melts into the bath and is filtered by the slag. As the melted electrode material fills the copper basin, it solidifies at a known rate. The crystallizer moves up the roll, or "goes upstairs," as chief engineer Blaskovits liked to say, leaving behind an overlay of new metal. So, XYZ had the crystallizer made in the United States and sent it to Czechoslovakia, where they used it to make resurfaced prototype rolls.

The next problem was having the rolls picked up and shipped. XYZ's usual forwarding agent had never dealt with Czechoslovakia or any other Communist country. New operating procedures were required. How to get to Bratislava? Where is Bratislava?

After much delay, the rolls arrived in Cincinnati and were installed in the Youngstown mill of LTV. The LTV mill promptly had a shutdown before the rolls could be fully tested. An indefinite delay in testing the "cold" rolls, rolls used for rolling cold sheet metal, such as is used for making car bodies, ensued. Rolls of higher chromium composition were also required from VUZ for testing as "hot rolls." Overlaid hot rolls would represent a new market opportunity for XYZ. If the resurfaced old rolls actually stood up in service as well as new forged rolls, then they could compete head-on.

The working surfaces of a metal object are "hardened" by a variety of techniques, all of which have the purpose of giving to the outer surface a hardness greater than that of the core material. Rolls in steel mills are subjected to tremendous stresses, and the depth to which rolls can be hardened influences the duration of their use or "campaign life." The longer a roll can be kept in service, the greater the productivity of a mill. Rolls that have been used awhile and are getting down to the limits of their depth of hardness can be "rehardened."

Electroslag resurfaced rolls, even if not equivalent to the new rolls in their service life, would offer a product superior to rehardened old rolls. The overlaid rolls would have a thicker "hardened" outer surface and longer campaign life than can now be produced by conventional technology.

Getting the electrodes produced for making hot rolls revealed much about the way the system works for VUZ or any institution in Czechoslovakia and other Socialist countries. Work associated with selling technology is an unpredictable event that is not part of an institution's research plan. All work for sample preparation, testing, and analysis is essentially "extra plan" activity that must be wedged into the schedule, often on weekends.

Foreign interest in a technology through active "selling" comes about in one of two ways: passive-active or passive-passive. VUZ has a technology and has given a description to Polytechna to offer as part of its portfolio of licenses. Polytechna, however, has hundreds of other topics to sell and generally is passive in its efforts, except for the occasional trade fair or the odd letter to an existing network of contacts. Employees at Polytechna are not rewarded according to the number of license agreements they conclude or income they generate. Their in-country representatives function like salaried postal clerks. They don't travel much. They don't knock on doors. Dynamic salesmen they are not. VUZ itself occasionally presents information at a scientific or technical conference one of its own researchers is attending. If interest is aroused, the person is sent to Polytechna for further discussion. But selling in the active, life-dependent sense of "sell or die" is not a fact of life for VUZ or any other organization in Czechoslovakia.

The passive-passive sale requires no promotion. It occurs when a potential foreign customer reads of some development in the technical literature and makes its own overtures to learn more about the technology. This is the "off-the-street-business" that comes from being known and being accessible. All but the most sophisticated American companies are unaware of the accessibility of Eastern bloc technology and tend to ignore opportunities when they learn about them.

The motivation for taking on the headaches caused by having an interested Western partner include opportunity to travel, prestige for the institute and its researchers, and money. Some portion of the revenues earned is typically divided among the scientists, their institute (in this case, VUZ), its ministry, and the commercial licensing organization, Polytechna.

Making to order twenty-foot-long bars of high-alloy-steel configuration is not easy in Czechoslovakia. There are not many companies that can do it in the United States, and those who can would have to be paid a tremendous premium. For VUZ to meet its extra-plan technology requirement for its American customer, it had to cajole a steel mill to take on unwanted extra-plan work to produce the electrodes. In a market economy, these same problems occur but are resolved, to a large degree, through the pricing mechanism. Being paid a premium in crowns does not do much for a producer whose performance is measured by meeting production goals. It is the old-boy network, personal relationships, and contacts that get nonplanned work accomplished. Only the personal contacts of the VUZ director and the system of favors owed eventually got the high-chromium electrodes produced for XYZ.

By the time the "hot" rolls were finally delivered to XYZ for testing, three years and two extensions on the option dates later, the company's financial condition had suffered from the economic problems of its major customer, U.S. Steel. Still a believer in the technology, nevertheless XYZ allowed the option to lapse, because it no longer had the luxury to spend funds on development projects unrelated to its most pressing operational needs.

The pushing and cajoling which VUZ undertook to complete its obligations to the U.S. firm have not been in vain. The American interest and willingness to pay money for the option provided the impetus VUZ needed to induce its domestic steel industry to take a second look at the technology, which was slowly molding away in its workshops.

The man who made VUZ a world leader in welding and surface treatment technologies is Vladimir Csabelka, now retired and in failing health. Trained as a metallurgist, he had spent much of the war working in a German occupation tank factory in Vamberg, Moravia, where he was recruited to work as an Allied spy. Important technical information about the novel material the Germans were using to armor their heavy tiger tank passed through a network stretching to Istanbul and London. When the war was over, Csabelka returned to Bratislava and served on the Commission for the Renewal of the Slovak Economy, the traditionally neglected half of the Czechoslovak economy. This became his base for obtaining funding in 1949 for his new metallurgical research institute, which would integrate several different disciplines. During the 1950s and 1960s, Csabelka's work was recognized internationally. From Cleveland's Lincoln Welding Foundation to Moscow's famous Bauman Technical University, his wide-ranging accomplishments were recognized with numerous awards. According to his son, Dusan, also a metallurgist, Csabelka regarded the institute as his family and his first family at that. He personally knew everyone at the six hundred–strong institute. He was a practitioner of what is called today "managing by walking around." Everyday he would walk throughout the institute, wanting to know what people were doing, what help they needed. If some were occasionally doing private work, using institute facilities, he did not object as long as they met their plan targets. And he remembered those who did not.

Out of the period of the 1950s came novel alloys for steel, a new spot welding machine for welding car doors which won international recognition in Brussels, and advances in welding cast material. Csabelka was also a great collector of young brainpower, recruiting the best of his former students from the Slovak Technical University, where he held a professorship in metallurgy. He was known for constantly quizzing his students and became famous for his spontaneous "car examinations" that would occur whenever he was driving with a student. Most of all, he was interested in the practical use of his students' knowledge and of the research being done at his institute. He was not merely interested in new materials but in how they could be made.

Csabelka's successes also brought him envy and enemies. In the late 1950s, under the Stalinist period of Novotny, Csabelka was increasingly attacked for being politically "wrong." In 1961, he was spared probable execution by being committed to a sanatorium by friends who declared him on the verge of collapse.

To be successful under socialism requires diplomatic and political skills to a much greater degree than in the West. Being good at what one does is not enough. Powerful friends and sponsors are also essential, along with a modicum of diplomacy and tact. In a system with no private sector, everything is political. And politics is a risky business, as Gustav Husák, the successor to Alexander Dubček, liked to say. He should know. Husák himself spent several years in jail for being politically wrong during the Stalinist period.

Cuba and Wichterle have not been torn down by the envious. Their international success has no doubt helped them. But they are also modest men who have learned to stretch the limits of the system with political skill. Wichterle did get his stripes torn off in 1969, but the consequences to his career were negligible. Afterward, he never baited the system publicly or sought to embarrass it. Independent-minded, yes. Foolishly antagonistic, no.

The Prague Spring now has become twenty years later the Moscow Spring. Roles are reversed. The more conservative Czech government is censoring Soviet TV broadcasts. Moscow is now the center of a bold reformist movement and sending shock waves through Eastern Europe and the world. The difference is that Moscow does not have to worry about Czech tanks rolling into Red Square.

The Czech leadership has not yet decided what it thinks about *perestroika*. Unlike that of the Soviet Union, its agricultural system works reasonably well and contributes to feeding other bloc countries. Compared to the Soviet Union, Czechoslovakia is the status quo–oriented country but, with its Wichterles and Cubas, carries the seeds of its own regeneration. Once banned for being too far ahead, the current leadership is waiting to see how permanent the Gorbachev revolution will be.

Chapter 3

The Red Baron

An institution is like a tune—it is not constituted by individual sounds, but the relationships between them.

PETER DRUCKER

In the minds of most people, East Germany, officially the German Democratic Republic (GDR), is represented by "The Wall." As a result of some Madison Avenue thinking, the East German government has sprung its athletes upon the world to create a more benign image of international competitive strength in the peaceful arts of sport. Every four years, their athletes dominate prime time television to remind viewers that there is more behind The Wall than sinister Volkspolizei.

Less visible is the strong competitive position of the GDR in high-technology areas such as optics, industrial automation, computer-controlled machine tools, and microelectronics. Zeiss Jena, of optics fame, was producing equipment for electron beam lithography in 1979 to fabricate submicron line widths on silicon wafers. This was two years before the Pentagon announced its big push to promote very large scale integration (VLSI) technology. VLSI was aimed at packing more circuitry onto a given piece of silicon "real estate," as the wafers are referred to in the industry, by reducing the width of the circuit. The Pentagon's goal was to produce commercial-scale submicron chip technology by 1988.

Advanced distortion-free East German photogrammetric cameras are used for making extremely precise measurements from photographs. Applications of the technology range from sizing coal inventories and croplands overflown by the IRS to aligning wing attach points on General Dynamics' F-16 before final assembly. The British Ministry of Defense buys high-resolution, multispectral space cameras from Zeiss Jena.

The industrial application of electron beam technology is a field in which East Germany is internationally preeminent. The topic is intimately associated with the name Manfred von Ardenne. The Red Baron, as he is known in the West German press, is a political anomaly whose life history and scientific accomplishments make him one of the most unusual R&D directors and creative personalities in this century. Though a household name in the GDR, he is virtually unknown outside German-speaking Europe.

Over the years that I visited East Germany, I had heard the name von Ardenne many times. He had been involved in top-secret Soviet research projects after the war. He was reported to have a private research institute in Dresden which was technologically prolific. In 1963, von Ardenne, the physicist, had turned his personal attention to problems of medicine. His pet projects are methods for treating cancer by using oxygen to activate immune responses and slowing the aging process by using a multistep form of oxygen therapy that he pioneered.

My plan to meet von Ardenne was carried out through the assistance of the East German Embassy in Washington. As I was known to the East Germans as a technology broker, I suggested that my trip be arranged through the auspices of their State Committee for Science and Technology in Berlin, rather than through journalistic channels. The approach

East Germany's "Red Baron," physicist and entrepreneur Manfred von Ardenne.

agreed upon, my visit to von Ardenne in fall 1987 fitted nicely with my plans to go to Berlin with a client interested in East German technology relating to plastics and polymers. As the former home of I. G. Farben, cobearer of the great tradition of German chemistry, East Germany remains one of the strongest Eastern bloc countries in the field of industrial chemistry.

By chance, some of the centers of excellence in polymer technology are also located in Dresden. Representatives from the Technical University in Dresden traveled to Berlin to give us an overview of their work in polymer chemistry. Dieter Lohmann, the scientific secretary of the chemical faculty, spoke fluent English and had spent a semester at the University of Massachusetts Polymer Center, where he learned the fine points of touch football and college town merrymaking.

The next morning, Lohmann met me in Dresden at the Hotel Bellevue, a former royal chancellery located on the bank of the Elbe, where it looks across the river to several restored masterpieces of Saxon architecture: the Renaissance-style Semper Opera House, designed by Gottfried Semper; the Zwinger Museum, former palace complex of seventeenth-century Saxon princes; and the Catholic Hofkirche, designated a cathedral by the Vatican after its restoration. Dresden is the capital of Saxony, an independent kingdom until annexed by Frederick the Great of Prussia in the eighteenth century. A century earlier, its king, Augustus the Strong, was also king of Poland, a job he fitted in while producing, according to local legend, three hundred bastard children. In an earlier period, the Saxon court had provided Martin Luther refuge from Catholic persecution and a place to translate the Bible into German. Today, Dresden is best known for its senseless destruction by the Allies in World War II.

Lohmann was kind enough to drive me to the baron's kingdom, known locally as the Plattleite, after the street which leads to the complex of buildings housing the Institute of Special Beam Research, as it is officially known. The road to the Plattleite winds up a wooded hill interspersed with clusters of shops and residences. At the top, there is a village where we turned on to the Plattleite Street leading to von Ardenne's headquarters. The buildings of von Ardenne's institute are indistinguishable from the other commercial and private residences. No "institutional" look is apparent, other than a guard house as we enter Zeppelin Street to Building 22.

My appointment was for 8:30 A.M. Dieter left me at the door of an elegant baroque villa at the end of Zeppelin Street facing out across the valley toward Dresden. Obviously, this was no ordinary administrative headquarters. I was clearly in the man's house. From the grand marble-floored foyer, I could look out through the adjoining conference room to the view beyond. It was a breathtaking sight—the city laid out below, bisected by the Elbe, and off to the south a glimpse of the white chalk cliffs known as the Saxon Alps. After announcing myself to his secretary, I took a seat below a grand staircase whose walls were covered with ornate gold-framed pictures of historical figures, some perhaps from the von Ardenne family. There was no mistaking, however, the enormous portrait of Albert Einstein that covered the main wall opposite me. After five minutes' contemplation of my surroundings, from behind I heard a "Herr Kiser?"

I turned to meet a tall, trim, slightly balding eighty-one-

This baroque villa in Dresden serves as von Ardenne's home and as headquarters of the Institute of Special Beam Research.

year-old Baron von Ardenne, nattily attired in a blue blazer and gray flannel pants. He led me into his paneled office, which overlooked the valley. This, too, was an office of a man working comfortably at home. As I later learned, the upper two floors were the living quarters for the von Ardenne family, including room for his two sons and their families.

R&D Philosophy

Von Ardenne characterizes his five-hundred-person institute as being like Battelle Memorial Institute, the huge research center in Columbus, Ohio, a comparison that would be flattering to Battelle. Von Ardenne claims that his institute's research output saves the GDR $300 million per year, thanks to the economies realized by more efficient processes introduced by its industrial partners. Von Ardenne attributes his institute's success to selecting appropriate subjects to work on, maintaining close relations with industry, and having a tightly knit team in which much authority is delegated.

A certain modus operandi has developed over the years. Each new research task is examined in light of the whole development cycle—research, manufacturing, and introduction—with a look-ahead period of two to ten years, depending on the project. It is not enough simply to produce a good research result: The manufacturing technology, ancillary equipment requirements, personnel needs, and material supply must be available within the necessary time frame. Only those research tasks are undertaken which the institute and its industrial partner agree upon in advance can be introduced, should the research effort be positive.

Through an interweaving of responsibilities and joint effort, tasks are designed so that both parties, researchers and end users, have a mutual interest in success. When difficulties arise, it is expected that the commitment to mutual success will be more important than formal written contracts and protocols.

The research orientation of the institute is overwhelmingly in the direction of production processes. Research and equipment development typically occur in parallel. This is a chicken and egg problem that can create management problems. Sometimes a new technology can only be developed through experimentation on production equipment which must be modified, adjusted, or experimented with. On the other hand, to create new, more efficient production equipment, new technological solutions must often be developed.

To prevent the problem of bureaucratic turf and compartmentalization of responsibility that can lead to a "that's not my problem" atmosphere, von Ardenne has developed the concept of "total responsibility," which spans research and development and requires that the person with the responsibility see the job to actual completion. Such an approach demands granting managers a considerable degree of authority, as well as support, from the top. Greater decision-making authority and risk taking also bring greater potential for failure.

The "project leader" has the job of ensuring the success of all his "theme leaders." Project leaders assume responsibility for general areas of technology and balance the resource needs of all the "theme leaders." The "theme leader" is the person on the spot who works directly with the industrial partner, heading up a multidisciplinary team consisting of members of the four main departments: Design, Technology, Computer Control Systems, and Workshop. Technology is the department which determines the optimal physical methods to employ to solve a problem, Design develops suitable equipment to carry out the chosen process, Workshop builds, and the Computer Control Systems develops the necessary control technology.

Underlying this management philosophy is the notion that every contract with industry must achieve success, because research projects are only developed after careful discussion with industry. The "project leader" is responsible for *everyone's* succeeding. He supervises, exhorts, coaches, and balances the needs of the "theme leader" teams. Success, in turn, breeds a high level of job satisfaction that compensates for the heavier work load at the institute than in the average East German workplace.

This attitude parallels a general one within the GDR toward economic reform which places them at a different end of the spectrum from Hungary and the Soviet Union. Rather than go along with the now fashionable notion of allowing poorly performing enterprises to fail, the GDR approach seems to be one of patient paternalism, working with the failing enterprises to make them succeed. In order that all may succeed a little, the dominant attitude embodies the dictum "Only he has a right to criticize who is also willing to help."

High-performance organizations, such as von Ardenne's, function more by natural selection than discipline. If the core group is highly professional and dedicated, then, to use his metaphor, new people tend to attach themselves to the existing core on the basis of natural attraction in the same way crystals grow. Those who do not measure up feel uncomfortable and depart on their own initiative. This process of voluntary "deselection" results in a turnover of about 10 percent per year.

"He is 70 percent businessman, 30 percent scientist," said one of von Ardenne's associates in accounting for the successes of the institute. This is reflected in von Ardenne's keen sense of the importance of combining scientific research with practical results. He is also a man with a well-developed public relations instinct. Off his study is a "PR" room filled with neat piles of articles, press releases, slides, tape recordings, and copies of his newly released autobiography published in West Germany.

Von Ardenne's favorite subject is his multistep oxygen therapy and related cancer research. Over the past twenty years, with the backing of the GDR government, von Ardenne has built up a medical research team over which he has taken personal command, leaving the electron physics to his trusted and talented deputy, Dr. Siegfried Schiller.

After filling me up with literature about his medical research, von Ardenne gave me a tour of the "house treasures." These were found in his magnificent conference room that looks out on the valley below. In the center was a round white table with gold leaf trim with matching Louis XV chairs. To the right of the entrance in a corner was the desk of Otto Warburg, a Rockefeller Foundation-supported Nobel Prize winning biochemist of the 1930s who put von Ardenne on the trail of cancer research with his exhortation "Go after the big unsolved problems of our time."

The Inventive Prodigy

Next to Warburg's small rolltop desk is a scientific artifact that represents one of von Ardenne's most significant pre–World War II achievements in the field of electron physics—the Raster, or scanning electron microscope. This invention, which von Ardenne was inspired to sketch out on a piece of paper in 1937, made it possible to use a beam of electrons to resolve an object down to 10 nanometers, or 1/100,000 meter, a significant improvement over the existing state of the art. The further development and application of this technology occupied much of von Ardenne's research energy through the end of World War II. Only twenty-five years later did an English firm bring out a similar product that had been developed by McMullen. For twenty-five years, his electron microscope represented the state of the art, leading to significant contracts with the electronics giant Siemens.

During the early war years, von Ardenne published the first book on electron hypermicroscopy, describing the technology he pioneered. Electron microscopes opened up new frontiers of biological research, enabling scientists to peer into cells

Von Ardenne beside the original Raster scanning electron microscope, which he invented in 1937.

and observe bacterial reproduction. This book appeared during the war years in Japan, the United States, and the Soviet Union. The Japanese were particularly influenced by it, and it provided their industry with a strong impetus toward the use of electronmicroscopy and electron beam devices. The Russians, as it turned out, were also close readers of von Ardenne's work.

In 1937, von Ardenne was a widely known prodigy in applied physics. By age twelve, von Ardenne had built his first radio receiver and by age sixteen had obtained his first patent for an improved method of sound selection for wireless transmission. When he was seventeen years old, von Ardenne was already self-supporting, thanks to his technical accomplishments, and paying his parents rent in order to help with the family budget. His father, Egmont, was a former military officer whose government salary was spread thin to raise five children. From 1924 onward, von Ardenne was earning enough money from his inventions and publications that he not only paid for his room and board at home but managed to continue his education and career without any further financial help from his family.

His most important early invention was the integrated electronic switch, which would reduce by one-third the cost of radios. It was sold to Siegmund Loewe, whose firm, Loewe-Radio, exported millions of the new receivers equipped with simpler and cheaper single tubes. Von Ardenne's success with Loewe, himself a physicist of high repute, taught him an important lesson. When von Ardenne described his work to Loewe, within minutes Loewe fully understood its implications and signed an agreement with von Ardenne in a matter of days. No internal rate of return was calculated; no market studies were made. As future experiences were to confirm for him, the more technically knowledgeable the decision maker, the faster a new technical concept will progress to the stage of practical application.

Another lesson von Ardenne was learning was that the

sources of technological progress were more and more frequently coming at the intersections of different disciplines. By his early twenties, von Ardenne already grasped the importance of multidisciplinary research. It was impossible to foresee in 1925 that such future disciplines as microelectronics, plasma physics, electro-optics, biology, and computer science would play a role in the success of his institute's work. In light of the continuing unpredictability of fields of study that would be important in the future, von Ardenne concluded that specialized study should be conducted mainly in connection with specific research tasks, or practical problem solving. For it is in actually trying to solve unfamiliar real-world problems that the most creative solutions are produced. The heart of the matter can frequently be grasped more readily by someone who sees a subject freshly. Formal education should be focused on the basic disciplines of natural science: mathematics, physics, and chemistry.

These were all early lessons and observations that von Ardenne was to put to use for the rest of his life. In addition to the triode and scanning electron microscope, he modified Braun's electron tube by developing an electrode for increasing 200-fold the intensity of the fluorescent spotlight. This invention, in turn, played an important role in the later introduction of cathode ray tubes in oscillography, radar, and television.

By the end of World War II, von Ardenne was deeply involved in Nazi research programs in his subterranean atomic research laboratory in Berlin. In addition to radar, his group was working on sixty-ton cyclotron accelerators and magnetic isotope separation for making radio isotopes. Fortunately for the world, the Nazi leadership after 1933 was indifferent to the importance of atomic physics and provided little support for research, nor were German physicists accustomed to taking time to beg the state for money.

On to Suchumi

In April 1945, the front reached Berlin-Lichterfelde. An opportunity was provided to von Ardenne to take his family and important documents with him to a location west of the Elbe. Sensing the direction of Germany's future, von Ardenne nevertheless chose to stay. Although his memoirs are not explicit, he knew he was throwing in his lot with the Soviets. A strong sense of loyalty to his team and the research "home" in Berlin-Lichterfelde that he had created over a twenty-year period was probably as important to his decision as his leftist political leanings. This refusal to abandon the ship and insistence on staying with his researchers continued a tradition of team solidarity that persisted far into the future. An immediate result of his decision to remain at his post was the rapid restoration of order at his institute. Windows were repaired, walls repainted, equipment unpacked and made operational—an oasis of quiet and orderliness quickly established itself amid the ruins of Berlin.

When the liaison officer to the Soviet Academy of Sciences, General Machniov, appeared at the von Ardenne research bunker in May 1945, he had with him a delegation of leading Soviet scientists: Artzimovitch, Flerov, Kikoin, and Migulin. Abraham Joffe, the dean of Soviet physics in Leningrad, had been following von Ardenne's work in the field of electron microscopy closely. The delegation had come to see what facilities existed and to learn more of von Ardenne's work.

At the end of the day's discussions, General Machniov handed von Ardenne an offer for scientific and technical cooperation, an offer he had anticipated, knowing from radio broadcasts that the Yalta Conference agreement had put German scientists at the disposal of the victors. Von Ardenne, however, had been told by his close colleague nuclear physicist F. G. Houterman that the Soviets attached great value to people who made significant contributions to the cause of science and technology. This was an assessment that proved to be true.

Nine days later, Machniov returned, this time with Colonel-General Saveniagen, a deputy minister president of the Soviet Union, and, later, the founder of Magnitogorsk, a new industrial complex for opening up Siberia's wealth. Saveniagen was to become von Ardenne's high-level guardian angel for the next ten years, but on that day of May 19, 1945, Saveniagen had come to deliver a proposal. The proposal was for the construction of a new technical physical research institute to be directed by von Ardenne. The topics of research were to be electron microscopy, various types of microanalysis using electron probes, isotope separation, and continuation of other research already in process at Lichterfelde. Von Ardenne accepted the proposal without hesitation.

Two days later, leaving their children behind, von Ardenne and his wife left for Moscow, where he was to sign the research agreement. Within hours of von Ardenne's departure for Moscow, the contents of his Lichterfelde research facility were being packed up by hundreds of Soviet soldiers. Seven hundred and fifty crates of equipment, sensitive instruments, and documents were packed, including the sixty-ton cyclotron, right down to the children's dolls and toys. A nearby bowling alley was ripped up to supply the wood necessary for the shipping crates.

Three weeks later, still with no negotiations, a car arrived at their dacha in the suburbs of Moscow. Hoping it was the representative of the government who would take them to the negotiations, the von Ardennes had the happy surprise of seeing instead their two-year-old son and five-year-old daughter and assorted other relatives. Along with the equipment, the Soviet packers had delivered von Ardenne's coworkers and families to a sanatorium near Moscow for rest.

By the end of June, von Ardenne was finally called for discussions and told that his future institute would be in the

Soviet Union, but he had a choice: Moscow, Crimea, or Georgia. On the theory that natural beauty was good for scientific stimulation, von Ardenne chose Georgia and asked for a location as close to the Black Sea as possible, a wish which was subsequently granted.

During this time, von Ardenne also heard that Gustav Hertz and his group were also coming to the USSR and were being set up with a similar institute. When von Ardenne learned of this, he persuaded the Soviet authorities that it would be mutually beneficial for Hertz's institute to be located near his. In the years ahead, the proximity of these two German institutes working side by side was a source not only of valuable scientific cross-fertilization but of mutual consultation over a host of personnel problems.

The location for his Technical Physical Research Institute was Suchumi, in Abkazia, an autonomous region of the Georgian Republic conveniently near the Black Sea. The main building of von Ardenne's institute was a former three-story, tile-roofed sanatorium consisting of one hundred rooms. It was located in a lush botanical garden of subtropical diversity and beauty consisting of banana trees, bamboo, cypress, cork bushes, camellia trees, and fig trees. The use of the building's space was being planned in accordance with the Lichtenfelde research program that had been agreed upon with General Saveniagen in Berlin.

August 6, 1945, changed all these calculations. Days after the bomb was dropped on Hiroshima, von Ardenne was informed by General Saveniagen that Marshal Beria now had different plans for the institute. The events in Japan would require von Ardenne to redirect his institute's research to help with the development of a Soviet atom bomb. Von Ardenne says that he had a strong aversion to weapons research and had spent much of his scientific career under the Nazis trying to avoid direct weapons research, turning down an offer by Wernher von Braun in 1941 to work on rockets. This time, von Ardenne's arguments were to no avail. Despite strong protests to Saveniagen, von Ardenne was told that the decision for him to work on bomb development was irrevocable.

By the middle of August, von Ardenne had to drop everything and go immediately into Moscow. On his way, he was informed that he was being taken to Marshal Beria's office, next to the infamous Lubjanka prison of the KGB. Von Ardenne describes the scene when he entered the conference room: Marshal Beria rose to greet him from the end of a long table surrounded by leading Soviet physicists: Kurchatov, Alichanov, Halperin, Artzimovitch. Von Ardenne was given a seat next to Beria, who began the meeting by announcing to him that the government of the Soviet Union wanted the development of the A-bomb to occur in the institute which he would be directing. Von Ardenne had about ten seconds to gather the words that would determine his fate.

"I consider your proposal to be a great honor," he responded, "for it shows an unusually great trust in my capabilities. The solution to the problem you have posed has two parts: (1) the building of the bomb itself, and (2) the development of an industrial scale process for the separation of uranium 235. The isotope separation is actually the most difficult problem of the two, the real bottleneck. I propose, therefore, that my institute has as its principal job the separation of the isotopes and that the leading Soviet physicists before me here carry out the development of the bomb itself as their contribution to the homeland."

Beria then broke up the meeting for a short consultation with the Soviet physicists. After a few minutes, Beria informed von Ardenne that they agreed with his proposal. A detailed discussion ensued in which von Ardenne expressed concern about the availability of the necessary resources to develop industrial-scale production facilities. As the American literature demonstrated, huge investments in costly facilities are required for producing the fissible uranium 235 isotope whose presence in nature is 1/10,000,000 that of the uranium 238 from which it must be extracted. Beria assured von Ardenne that the Soviet government would make available the necessary resources. After the meeting, further working sessions to plan an accelerated research program took place. Years later, von Ardenne concluded that the safest place to be is in the mouth of the lion, preferably in a hollow tooth. When he had returned to East Germany, von Ardenne was once introduced by President Grotewohl to Khrushchev with a reference to his ten years in Suchumi. Khrushchev's eyebrows went up: "Ah, you're the Ardenne who so cleverly got his head out of the noose."

To staff his institute with the necessary specialists, von Ardenne had to draw on the German prisoner of war population. The Russians sent down to Suchumi from Moscow those prisoners with a scientific and technical background. Organizing a series of interviews lasting no more than five minutes, von Ardenne had the Radamathanian task of deciding the fate of the almost-dead, for many of the prisoners were suffering from extreme malnutrition and disease. Yet mistakes in the selection of the technically critical personnel to run the institute could mean failure for von Ardenne's mission.

Selection to serve on von Ardenne's research team was a short-term blessing for prisoners suffering under poor conditions, but required staying at least six more years—which turned into nine—in the USSR. In the matter of morale and the role of "human factors," General Saveniagen showed himself to be extremely understanding and helpful. In October 1945, von Ardenne spent a whole night with Saveniagen in his special railroad car discussing the family needs of the German workers. They decided that all the prisoners who were selected for work in Suchumi would have their families brought to stay with them. Both men realized that in times of stress, family support was essential.

For the next nine years, von Ardenne and his Germans worked under conditions of virtual confinement to their research facilities and were accompanied at all times by security personnel, whether grocery shopping or picnicking in the mountains. Technical progress toward solving the problem of separating the uranium 235 isotope was accomplished through an administrative mechanism which von Ardenne was to duplicate in the GDR. The main instrument of planning and decision making for this highly complex undertaking was the technical soviet, or council, which met once a year. The technical soviet became a significant administrative tool, because it brought together in one place and in one marathon meeting the leading figures in the scientific community, government, and industry with a common interest in the success of the project. At this annual "sitting," important decisions for the coming year were made by people with the power to execute those decisions and with leading industrial partners who had to implement them. It is to this Soviet management tool that von Ardenne attributed the relatively rapid progress on the A-bomb.

During the next four years, Institute A, as von Ardenne's Germans were referred to, made enough progress that by 1949 von Ardenne had to move up to Leningrad for an extended period for industrial-scale pilot testing. In Leningrad, the huge two-hundred-ton electromagnets were being made at the Electrosilia Works. This technology provided the basis for the later pioneering Soviet work that used magnetic fields to contain superheated plasma needed to produce energy by atomic fusion. These magnetic rings replaced conventional physical containment walls, which could melt under the high temperatures being created by the plasmas. Plasmas are nothing more than man's attempt to reproduce the effects of the largest plasma reactor of all—the sun, a large ball of ionized matter. Plasmas hot enough to fuse atoms are minisuns. This magnetic containment concept of achieving fusion power, which has been adapted by American nuclear physicists at Princeton University, had its beginnings in the Lichterfelde Institute in Berlin that the vagaries of war transplanted to a subtropical cypress garden in Soviet Georgia, and reduced to practice in the Electrosilia Works of Leningrad, itself a pre–World War I German electrical works owned by the Munich giant Siemens.

Initiatives and Continuity

By 1950, enough progress had been made on the practical solution of separating uranium 235 plasma columns contained in magnetic fields that von Ardenne began unauthorized work on new related topics of interest to him and high future value to the Soviets. These topics included the development of high-energy ion sources, later to be known as *duoplasmatrons* in the technical literature, and precision electron beam oscillographs with extremely high resolutions. The duoplasmatron ion generator that von Ardenne started developing in 1950 produced a virtual 100 percent utilization of the ions. The ability to produce high-energy charged particles efficiently found a wide application in atomic physics for particle accelerators and for ion implantation, which made possible the production of superthin coatings for enhancing the service life of cutting tools and industrial componentry subject to high abrasion and wear. Practical interest in the use of ion source propulsion systems for space flight was indicated by von Ardenne's selection in 1965 for membership in the International Astronautical Academy in Paris.

Another instrument that von Ardenne developed outside the research plan was a precision electron beam oscillograph. This, too, was a first: an instrument which displayed a given voltage level and wave form by the deflection of electrons. When these unplanned results were made known, considerable displeasure was expressed by those concerned that the planned research activities were being prejudiced. But when the significance of the work was realized by representatives at the technical soviet and data supporting the claims of the duoplasmatron's efficiency were provided, criticism dissolved into praise for von Ardenne's initiative. He was singled out for emulation by other Soviet research directors. As a result of his successful initiative, von Ardenne was given a virtual free hand to pursue what he wanted during the rest of his stay in the USSR.

One of von Ardenne's main concerns as a research director was maintenance of scientific continuity, as his research team moved from place to place. This concern was part of a larger aim of von Ardenne's going back to the 1930s, of regularly organizing and systematizing the results of his scientific work by putting it in easily graspable and retrievable form—what he called his *Wissenspeicher*, or "knowledge bank." By 1951, von Ardenne was already beginning to plan his return to Germany. This aim was to keep the important knowledge gained from the prewar Lichterfelde period intact as well as to add to it the new scientific and technical insights gained at Suchumi.

Starting in 1951, von Ardenne used every free hour to organize into tables and charts the complete knowledge gained from the different fields of activity being pursued by his group, a kind of scientific "concentrate," as he called it. One purpose behind this effort, aside from preserving knowledge in an easily retrievable form, was helping industrial researchers to access the core knowledge and art in specialized fields outside their own area of expertise. Maximizing intellectual productivity was a problem which preoccupied von Ardenne. He was convinced from his experience in electron physics that almost 50 percent of the time used to solve research problems was actually spent looking for once-read and forgotten articles, searching for data, formulas, material properties, design concepts once seen somewhere, not to speak of simply reinventing the wheel. To produce a "concentrate" of knowledge in applied physics, von Ardenne used the combined tools of data charts, tables, formulas, and illustrations of design concepts, all with relevant literature citations and excerpts, organized in an encyclopedia format. In this way, von Ardenne also saw the possibility of systematically transferring the valuable ten years of work done at Suchumi to his new institute in East Germany.

Never one to waste time foolishly, von Ardenne had sought out the permission of Professor Jemeljanov, one of the leading Soviet physicists, before compiling his scientific knowledge in a manuscript which could be taken back to the GDR. Not only was this request granted, but shortly after his return to the GDR, von Ardenne was notified by the Soviet Embassy that he would also be receiving a collection of the drawings and designs relating to top-secret devices that had been developed at his institute in Suchumi. Together, these materials allowed a seamless web of knowledge to be transferred back to East Germany.

This information was published in 1962 under the easy-to-overlook title *Tables for Applied Physics*, consisting of twenty-five hundred pages of interdisciplinary "concentrate" in a three-volume recipe book, a book which generated letters from American physicists who wondered how they had gotten along without it for all these years.

The fear that von Ardenne took with him the day he left Marshal Beria's office in 1945, after agreeing to work on the problem of uranium 235 isotope separation, was made vivid in the story he had fortuitously read shortly before the meeting. It was a story that happily did not repeat itself in the realm of physics, of the construction of St. Basil's Cathedral in Red Square. Legend has it that Ivan the Terrible considered the building so magnificent that he had the two architects' eyes gouged out to be sure they would never build another as beautiful.

As early as 1950, von Ardenne had already decided to exercise his choice to return to East, rather than West, Germany, though it is not clear how much choice he really had. In his memoirs and in conversation, von Ardenne gives the impression that he did have a choice, which, on ideological grounds, he exercised in favor of East Germany, despite the fact that his former Lichtenfelde institute was located in West

Berlin. This consisted of four reconstructed buildings that would have still been his property, had he returned. However, he admits that his selection of Dresden for his institute was a disadvantage for maintaining his team, as staying with von Ardenne meant committing oneself to live under communism. Many of his fellow researchers nevertheless stayed on with von Ardenne and formed the core of his new institute.

The decision to relocate in East Germany was also influenced by financial convenience. During their ten-year stay in the USSR, von Ardenne and the other Germans drew a salary, most of which was deposited directly into an East German bank account at the favorable exchange rate of two marks to the ruble. Furthermore, von Ardenne had received the Stalin Prize in 1953 and other forms of recognition by the Soviet government which yielded financial gain, accumulating by 1955 a rather tidy savings account. It was this Soviet "bomb" account which provided von Ardenne the means to buy property in the area known as the White Deer, on the outskirts of Dresden.

The other decision which made life easier for von Ardenne was the willingness of the Soviet authorities to recognize that virtually all of the significant equipment that was dismantled in Lichtenfelde in 1945 and taken to Suchumi was von Ardenne's personal property and would remain so. The patents underlying all these devices were in von Ardenne's name, and he had built them with his own institute's money.

When I asked von Ardenne about his decision to live in East Germany, he did not make any direct allusion to ideology, though it is clear from reading his autobiography that his nascent sympathies toward socialism grew during his stay in the USSR. The answer took the form of the observation that most of the leading German scientists, especially those with a commercial-minded orientation, returned to the Federal Republic of Germany after the war. He thought he could have more impact in the GDR, where there were few businessmen-scientists. It almost sounded as if he'd rather be a big fish in a small pond, yet he had already demonstrated that he was capable of distinguishing himself in prewar Germany. Lurking in his divided German soul—one part socialist, one part businessman, one part scientist—may also be a genuine nationalist who wanted to help that part of his country which would benefit more from his abilities.

Another reason von Ardenne may have a benign view of socialism is that his experience is atypical. As a high-achieving foreigner granted exceptional status in the Soviet scientific establishment, he had all the privileges that special status can bestow, despite technically being a prisoner-of-war, one who was commandant of his own very special scientific POW camp.

Von Ardenne's career in the Soviet and East German scientific worlds illustrates an anomaly of Soviet-style systems, though the tradition also has general European origins. Unlike in the United States, great prestige there is attached to pure intellectual life. Like the "Herr Professor" of Germany in earlier days, the "Academician" is a figure of tremendous status in Communist systems, part of a special elite with privileges that set him apart. Incongruously, the societies which place the greatest value on scientific and technical achievement as badges of social status have the greatest difficulty in implementing those achievements in the everyday world. Partly for this reason, Communist economies tend not to rely on the system—which doesn't naturally encourage innovation—but to look to individuals and bestow upon them special resources and extraordinary conditions to help them overcome the system's own inertia. Von Ardenne, a largely self-taught, uncredentialed genius, produced one of those islands of real performance which the system recognizes and nurtures, sometimes with ambivalence.

In a 1963 meeting with Walter Ulbricht, then first party secretary, to press for more rapid introduction of his institute's achievements, von Ardenne recalls Ulbricht's saying, "Did you know, I helped you a lot before you returned to the

GDR? We thought back then that you were a remarkable and strange person. You wanted to work with capitalist methods in a socialist system. We recognized, however, that your efforts would be useful for the cause of progress."

Von Ardenne has maintained his privileged status in the GDR precisely because he has been an achiever, even without academic credentials: he had only four semesters of university-level training. The institute under his direction has been tremendously significant for the GDR economy. He has been a successful maverick, and his successes in the Soviet Union have not hurt him either.

A strong believer in concentration of force, von Ardenne influenced the GDR's scientific research plan early when he helped persuade the government to focus its research expenditures in a few critical areas of industrial development and to achieve world-class results in those selected fields. Of course, research has to be combined with volume production, quickly put into practice. To gain a commercial advantage, von Ardenne argued that speed of introduction into the marketplace was far more important than patent protection to gain a place in the world market.

To aid in rapid commercialization of research, von Ardenne pushed for an administrative mechanism similar to what he had experienced in the Soviet Union: assigning leading industrial scientists with responsibility for commercializing key research results to the scientific executive council, giving them direct access to the highest government officials. Von Ardenne also argued for accountability of scientific research to show results commensurate with the investment. Where this did not occur, changes should be made. Just as, in agriculture, one lays down more fertilizer on those areas which bear the most fruit and spares it on the barren parts, so with scientific research. This point of view was predictably controversial.

The Research Institute von Ardenne has embodied the life and principles of its founder. Von Ardenne does not talk about management; he talks about maintaining "tempo," which is achieved through a combination of common sense, Ben Franklinesque principles which might have come from a research director's *Poor Richard's Almanac,* and practical experience.

As early as 1938, von Ardenne had identified the requirements for maintaining "tempo." These included the following nuts-and-bolts prescriptions: Don't put off to tomorrow what can easily be done today; answer letters immediately; keep a well-stocked inventory of standardized parts and components; pay bills quickly. On a higher level, von Ardenne advocated maintenance of direct contacts with leading international specialists to keep informed of developments abroad; construction of laboratory test facilities as similar as possible to actual factory conditions; systematic organization and reduction of essential scientific and technical information to tables, formulas, and charts for use by researchers; flexible and dynamic organizational structure; and, of course, rapid transfer of the results to practice.

The Institute of Special Beam Research has produced results which match von Ardenne's words. Its output is impressive, showing the GDR to be a valuable source of high technology for the USSR, a connection which has been largely ignored by the U.S. government export control community. In many "sensitive" areas of technology, the von Ardenne Institute has played a key role in putting the GDR on the forefront of advanced materials processing technology or has been *the* recognized pioneer. An example of "tempo" is provided by the introduction of electron beam technology for the industrial refining of metals.

In January 1959, von Ardenne developed the idea that specialty steels and reactive metals could be melted in a vacuum furnace, degassed, and purified to a high level by using powerful electron beams (E beams). Nine months later, in October 1959, the first multichamber E beam furnace was put into operation. Von Ardenne attributes this achievement to the use of the technical soviet management model, incorporating a high-level, superinstitutional body of leading personalities from the different producer-user organizations.

Over the next four years, high-performance electron beam guns, representing the "technology" end of the system, were exported to Japan, India, the Soviet Union, and the United States. Large furnaces, fifty feet high, which could melt twenty-ton ingots, were constructed for VEB Hans Beimler in Hennigsdorf, East Germany. Using an E beam energy source, these huge ingots were remelted and further purified for use in high-performance applications such as jet engines. In such applications, even small amounts of impurities, such as sulfur, can lead to failure. E beam can reduce impurities to inconsequential levels, thereby improving overall strength. With this technology, the GDR was spared the expenditure of hard currency for importing expensive specialty steels and actually became an exporter of such metals, as well as of furnaces.

The development of the electron guns for melting furnaces represented the birth date for von Ardenne's thrust into the general field of electron beam technology. By 1964, his achievements had won him an invitation to be the opening speaker in Toronto for the World Conference on Electron Beam Technology.

Thanks to the experience with E beam machining already developed in 1953 in the USSR, in the mid-1960s von Ardenne was building the first computer-controlled electron beam micromachining equipment for making circuits on electronic materials. By 1971, production units for micromachining were installed at the Ceramics Works in Hermsdorf. In 1965, von Ardenne had put into industrial use the first electron beam welding installation for ultrathin high-quality welds. Welding is like a form of surgery, and, as in surgery, it is desirable to keep as many of the potential impurities in the air as possible out of the wound. As air contains impurities that can "infect" the weld, E beam is used in a vacuum, preventing ambient impurities from entering.

Von Ardenne was also active in related areas of technology. The year 1963 saw the development of original plasma torches consisting of a one-millimeter-thin, 10,000° Celsius beam of superheated energy. This technology made possible rapid cutting of very hard and brittle specialty steels with high temperature resistance. After a demonstration of the technology on GDR television, hundreds of orders came in from East German metalworking firms and shipyards. Over one-half of the eight hundred plasma torches made by the Mansfeld Kombinat were exported to Japan for use in modernizing their shipyards. Only in 1981 did an American firm learn about the East German technology.

Von Ardenne's Institute of Special Beam Research works with over fifty different industrial partners in the GDR, ranging from producers of thin film vacuum deposition equipment (Hochvacuum Dresden) and magnetic isotope separators (RFT Messelectronic) to manufacturers of state-of-the-art microelectronic production equipment (Zeiss Jena). Nor is there any reason to assume that the Soviet Union does not have access to technology it wants from the GDR, through either direct commercial channels or scientific and technical cooperation. Von Ardenne's group maintains close connections with leading figures in the Soviet scientific community. What is true in von Ardenne's case holds in varying degrees with other fields of technology in optics, flexible manufacturing, robotics, and other subjects considered sensitive by the U.S. government. The East German–Soviet technical establishments are warf and woof.

East Germany is a significant industrial force and technological power in those areas in which it has chosen to concentrate. Just as Western businessmen usually overlook the Soviet bloc as a source of new technological developments, so U.S. export controllers have not been accustomed to thinking about Eastern Europe as an alternate source of advanced technology for the Soviet Union.

After writing an op-ed piece in *The Washington Post* in 1984, I received a visit from a captain in the Department of Defense (DOD) Office of Export Control who was intrigued by the idea that there might actually be sources of advanced technology in Eastern Europe comparable to those in the West that the Soviets could access.

One reason Eastern bloc organizations like to buy from the West is that buying represents an opportunity to go on a pleasant shopping spree—to travel, see some sights, buy something nice that cannot be easily obtained at home. Access to travel is the highest of premiums to be awarded in the East. What normal, travel-starved Communist would rather go to Leipzig than Hamburg, Minsk rather than Paris? Obviously, the best technology is in the West.

The problem became so acute in East Germany, whose experts always looked automatically to West Germany to satisfy their technological needs, that in the early 1980s a policy had to be instituted to require that East German organizations wishing to buy technology from the West demonstrate first that the desired technology was not actually available in the USSR or another Eastern European country. Millions of dollars of hard currency have been saved by this administratively sound policy, which has deprived untold East German households of the latest in Western VCRs.

East Germany has one of the largest industrial bases in the world, despite its small population of seventeen million inhabitants. It has the least amount of Western debt of the Soviet bloc countries, though it accepts much largess from West Germany in the form of subsidies and grants. Unlike in Hungary, reforms in East Germany have been overlooked in the Western press. Today's Prussian rump state has not jumped on the bandwagon of free market mechanisms, choosing a system of decentralizing industrial management and lodging great authority and autonomy to its *Kombinats*, a model not unnoticed by those Russians who prefer administrative approaches to market mechanisms.

The Wall notwithstanding, East Germany represents historically the geographic and philosophical heartland of German socialism and some might say of modern factory state socialism generally.* Not only does East Germany not believe it needs *perestroika* it surely considers itself to be more tutor than student of reform socialism. At a time when the Soviets are asking fundamental questions about the essence of Marxist socialism, they might turn to the father of revisionist socialism, the Berlin-born Edward Bernstein, who challenged Marx almost one hundred years ago. The problem for Bernstein during the period of pre–World War I German prosperity was similar to that faced by Gorbachev today. The predictions of Karl Marx, based on his scientific method of historical analysis, did not seem to find much empirical support in the Germany of the 1890s, where class alienation and concentration of wealth were lessening rather than growing. For the Soviet Party leadership, the problem with the Marxist model is the opposite. Rather than providing abundance, socialism is producing scarcity; instead of class harmony, it is creating social tension and privileged elites.

Further, Bernstein rejected the implication of moral neutrality implied in Marx's claim to have a scientific method. No ism, said Bernstein, is free of bias. Socialism is a system of demands. In Marxism, these demands are clothed in terms of predictions, but this is merely scientific camouflage, disguising a particular concept of a just society. His desire for equal partnership in society is not scientific, but an ethical judgment of the social interest.

---

*In his *Road to Serfdom* (1944), Friedrich Hayek notes that French socialists openly recognized a kinship between socialism and the organization of the Prussian state. A certain brand of socialism was to be inspired by the ideal of running the whole state on the same principles as a single factory. Berlin was the breeding ground for many important leaders of the nineteenth-century German socialist movement, among them August Bebel, Karl Liebknecht, and Ferdinand Lassalle.

Gorbachev has now over seventy years of experience with the ethical judgments of Marxist socialism. Just as reality forced German revisionists of the early twentieth century to reevaluate Marx, now a different reality is forcing the Soviet Communist Party to rethink ethical tradeoffs in the pursuit of what "ought to be." And today East Germany, with its more efficient economy and special relationship with West Germany, has a different notion than the Soviet Union of what "ought to be."

Chapter 4

Thorns in the Side

Inequality is the cause of
all local movements.

LEONARDO DA VINCI

"*Perestroika* is a mental as well as an economic process. Everyone should subject themselves to *perestroika*. You can't pat yourself on the back for long, or you die." Vladimir Kabaidze is one of the USSR's most successful industrial managers and a strong supporter of Gorbachev's efforts to reform the economy. "Old ways die hard," Kabaidze was explaining during our outing to Mount Vernon. He was one of two Soviet "industrialists" to be included in the annual Dartmouth Conference, a back-channel forum of Soviet-American dialogue that has been going on since 1960. In spring 1988, the Dartmouth Conference had been held in Austin, Texas, and I had arranged to intercept him on his way back to Moscow via Washington.

"The prohibition against drinking that Gorbachev introduced is typical of the old ways. It is against the spirit of *glasnost* and *perestroika* to impose this decree upon the people. This approach to the problem is part of the old mentality of the command economy. Of course people should not be allowed to drink on the job—that is perfectly reasonable. But to severely restrict people's access to alcohol and force abstinence upon people will not work—your Prohibition showed that—and this measure denies people choices that should be theirs." Kabaidze exemplifies the new way of thinking that Gorbachev is trying to promote.

The need to loosen the harness on the vital energy and creative spirit of the Soviet people without creating a runaway political nightmare that will overturn the central role of the Communist party represents a challenge requiring the highest order of statesmanship. Just as non-Russians, particularly Jews and Poles, played a vital role in the early success of the Bolshevik revolution, so Soviet citizens of Baltic, Armenian, and Georgian blood are the likely dynamos of economic *perestroika*, as people especially endowed with a spirit of commerce and entrepreneurship. The three personalities described in the following pages were selected because they are recognized achievers who employed atypical approaches, yet they also come from different areas of professional life.

The stories of eye surgeon Svyatoslav Fyodorov, agroindustrialist Albert Kauls, and machine tool entrepreneur Vladimir Kabaidze illustrate the dormant talent and energy which must be unleashed a millionfold for *perestroika* to produce significant results. Gorbachev needs models of what has worked, and little new has been functioning long enough under *perestroika* to provide guidance. He must look, therefore, to those who have already succeeded in creating mini-*perestroikas* under the Old Regime, now referred to officially as "the era of stagnation."

In a system which is constantly looking for examples of what works, these individuals have an ambiguous message for the Soviet leadership. Despite a political system which for two generations has practiced economic leveling and suffocated creative spirits, innovative individuals still exist: hardheaded, stubborn, contrarian, willful pushers who do not take no for an answer. This is the good news for Gorbachev. However, for the impact of such people to be felt within the society at large, there must be a critical mass of them. And to succeed in the pursuit of their creative vision, people must be able to be only "normally" exceptional and not supermen with powerful friends.

The Bulls of Production

Vladimir Kabaidze likes to say that he started *perestroika* fifteen years ago at his Ivanovo Machine Tool Amalgamation, two hundred miles from Moscow. Today Ivanovo comprises twenty-five thousand employees, including its joint ventures with Bulgarian and North Korean counterparts. It is the Soviet Union's only producer of advanced flexible manufacturing systems (FMSs), machining lines with individual reprogrammable modules that can quickly and automatically change tooling and cutting rates to produce a different type of part. Such systems are ideal for factories which must produce a wide variety of different components. In time of war, factories with FMSs would not have to be retooled to produce parts for tanks, merely reprogrammed. Ivanovo now has two new plants under construction which will add another ten thousand people to its work force.

As a result of Ivanovo's success in world markets, Kabaidze's enterprise was the first in the USSR to obtain direct foreign trade rights under Gorbachev in 1987. Unlike most Soviet producers, who must conduct their foreign trade activities through the Ministry of Foreign Trade and their own specific ministerial authorities, the Ivanovo Amalgamation is an autonomous commercial entity. Kabaidze can sign a contract anywhere in the world without approvals from other administrative bodies. Ivanovo was the first industrial enterprise to have its own hard-currency account, retaining 50 percent of the hard-won earnings from its exports to Japan, West Germany, Switzerland, Holland, Austria, and other countries. To service its foreign customers efficiently, Ivanovo has its own export department as an integral part of its business operations, just as General Electric or Siemens has. The department maintains two service people in Switzerland to provide service within twenty-four hours for European customers, who include leading machine tool manufacturers, such as Traub, Studer in West Germany and Naxos in Switzerland. In the early 1970s, Soviet manufacturing organizations rarely sold in the West, yet Kabaidze made it a matter of principle for his Ivanovo Amalgamation to produce for export in the toughest of all markets, the international one. "Export is the most cruel form of competition. Only through struggle do you get strong. The Soviet market is too easy; you can sell anything. If people don't struggle, they die. This is life."

It was this unusual tough-competition-is-good-for-you attitude, along with a general maverick reputation, that got Kabaidze an invitation to meet Gorbachev only ten days after Gorbachev came to power. He has remained an important adviser ever since.

The "struggle or die" philosophy of Vladimir Kabaidze, which has served him well in his fight against bureaucracy, sloth, and indifference, has deep roots in the painful experience of World War II. Struggle of the most existential nature began in 1942 when a seventeen-year-old Kabaidze went into the infantry to fight the Germans. Three times wounded, Kabaidze, four years later, was one of two surviving soldiers from his original unit, that had been through Stalingrad, Kursk, and the battle of Berlin. Kabaidze estimated his chances of survival at well below 1 percent. It is in these war years that he learned the meaning of struggle. They are years he prefers not to talk about, but it is clear that he considers the war a training ground that made the battles with Soviet bureaucracy easy by comparison.

Combined with his warrior toughness is a gentle, kind,

emotional nature. Kabaidze becomes particularly emotional when he talks about the meeting at the Elbe River with the Americans. His reminder of those happy days of celebration was the faded, illegible name *Frank Parent* scribbled on a piece of military script. In May 1988, Kabaidze cried like a baby when, forty-three years later, he met the man with whom he had spent two days of nonstop singing, drinking, embracing, and friendship. Barney Oldfield, a Litton Industries representative, had helped track down Parent after meeting Kabaidze during a visit to Ivanovo in 1987 as part of a delegation to discuss technological cooperation. Oldfield orchestrated the emotional surprise reunion one year later at Disneyland City Hall with Mickey Mouse benignly looking on—the result of nine months of detective work that led him to the retired seventy-two-year-old geophysicist living in Galveston, Texas, who had shared with Kabaidze on the Elbe River the overwhelming joys of victory. "Like all Georgians, he is very sentimental and emotional, particularly about his ties with the Americans he met at the Elbe river," observed Oldfield.

Born of a Georgian father, a mountain man who had made a career in the military, and a Russian mother, Kabaidze grew up as an army brat, moving constantly. Except for a few choice phrases, Kabaidze speaks no Georgian. Nor does he attribute to his parents the slightest influence on his interest in machine tools. Love of machines and building things apparently came from some recessive genetic strain. As a boy, he especially loved building boats, but it was during his two years in an aircraft works as a teenager that he saw his first metal cutting machines and developed a lifelong fascination for them.

After the war, Kabaidze's superiors tried to persuade him to make a career in the military, but he had had enough. First Lieutenant Kabaidze returned to finish high school, receiving upon graduation a "certificate of maturity." He then entered the Institute of Machine Tool Building in Moscow, where he

Soviet industrial manager Vladimir Kabaidze (left) reunited with Frank Parent (right). The two men first met as soldiers 44 years ago when Russian and American forces linked up in Germany. The surprise meeting was arranged by Barney Oldfield (center).

specialized in machine tool design. Upon graduation from the Institute of Machine Tool Building in 1952, Kabaidze found a job as a design engineer at a machine tool factory in Ryazan. The decision to go to Ryazan was motivated by a desire to build his own career "from a blank sheet of paper," rather than to stay in Moscow, where his wife had influential connections. At Ryazan, he moved quickly up the ranks to become the technical director of the enterprise within six years.

During his period as technical director, Kabaidze had opportunities to travel to Western Europe. In 1966, he took a trip which strongly influenced his thinking and his subsequent career. In Belgium, Austria, and Sweden, he saw the first computer-controlled machine tools in operation and was deeply impressed by their versatility and productivity-enhancing capability. He also learned about the importance that the Western producers attached to their customers and the extent to which they served their needs. Kabaidze returned to Ryazan a champion of computer controls. The general director of the works was predictably less enthusiastic, seeing more problems than opportunities. Chafing under the conservatism of his boss, in 1970 Kabaidze happily took an offer to direct the turnaround of a dismally performing new machine tool works in Ivanovo, two hundred miles north of Moscow.

Ivanovo was reputed to be dangerous for a man to walk the streets at night, lest he be attacked by love-starved heroines of socialist labor. As a center of light industry, in the postwar USSR Ivanovo was a city of women's work, depopulated by men who went off to work in heavy industry. Kabaidze's wife would have none of her husband's going off to Ivanovo to do "advance work" in finding an apartment for the two of them, rather than both moving and living in a hotel until suitable quarters could be found. Even though Mila Kabaidze had a good job as a doctor in Ryazan, a slightly disappointed Kabaidze found his wife settled into the hotel after he had spent only one uneventful night walking the dangerous streets of Ivanovo unscathed.

The plant that Kabaidze had been asked to direct had been marked for light, simple machine tools. It was a five-year-old copy of a Leningrad machine tool works, but the latter was designated for making heavy and complex tools, such as multipurpose machining centers. Kabaidze wanted to reverse this conceptual scheme and make Ivanovo a center for the most advanced, complex heavy-duty machine tool construction, using the latest in computer-controlled technology. Kabaidze was now a self-taught computer control expert and thought that a new plant would be the most appropriate for introducing advanced technology.

His proposal met howls of disapproval and counterarguments. The ministries had already decided the matter. Further, it was a crazy idea, as the basic infrastructure needed for sophisticated machine tool construction did not exist. There was no electrical or computer-based industry in Ivanovo to support the kind of control technology needed. Where would the people come from? There were no traditions to support this approach. All their arguments, Kabaidze readily admits, were objective and logical. However, they all failed to take one very important element of success into account—the will to make it succeed. Kabaidze had the will and desire to build a showpiece. Many of the older people at Ivanovo shared the thinking of the bureaucrats who said it was impossible. They, too, had to go. "Traditions are good for evolutionary change, but not for revolutionary change," observed Kabaidze, who swept away tradition in one fell swoop by bringing into his employ young people who had enthusiasm and who did not know what they could not do. Thus began in the early 1970s the repopulation of the "city of women" by Kabaidze's young bulls.

Success by Breaking the Rules

"He's an innovator who does not follow orders. He does things his own way," says Sergei Rogov, of the USA and Canada Institute in Moscow, who has studied Kabaidze's career. Beginning with contravening the plan for the new Ivanovo plant he was chosen to direct in 1970, Kabaidze systematically ignored the nostrums of the planning system.

After getting rid of people who represented "old thinking" and bringing in young world-beaters, Kabaidze started shaking new bureaucratic trees. He did not make products specified by the plan unless they were needed by customers. Nor would he accept Soviet-supplied products if they were not up to world standards. Kabaidze was accused of being unpatriotic and traitorous because he preferred to buy Western equipment, especially the electronics for the computer controls. Today, Kabaidze spends $60 million per year to buy the best of foreign-made components for his machine tools. Achieving this capability in opposition to his ministerial bosses came from Kabaidze's successes in implementing a revolutionary view of how his business should be conducted.

"The customer is God," exclaimed Kabaidze over shrimp snacks at the L'Enfant Plaza cocktail lounge, where I had my introduction to "The World According to Kabaidze." "My success came because I thought horizontally, not vertically. I worked to make my customers satisfied, not my bureaucratic bosses. In the end, it was my customers who protected me." To get hard-currency in order to provide the best electrical systems for the machine tools built at Ivanovo, Kabaidze went to his customers with brochures of accessories which he could supply on his machine tools. His customers, many defense industry plants, took some of their hard-currency allocations and gave them to Kabaidze, who then purchased the desired foreign components without getting an allocation from his own recalcitrant Ministry of Machine Tools and Tooling.

Even more revolutionary, Kabaidze went to his customers and asked them what their problems were. By offering to customers Ivanovo's willingness to tackle their most difficult production problems, Kabaidze developed a reputation as a supplier who cared and made a difference—not typical of Soviet suppliers, who are accustomed to produce only what is required and to expect the customer to be happy to receive anything at all. Service is a foreign concept in an economy chronically short of goods and materials.

It was service provided to the director of the Zil Engine Amalgamation that made Valery Saiken, now mayor of Moscow, a supporter of Kabaidze in the bureaucratic wars. When Saiken began the job over ten years ago, he wanted to convert engine production from gasoline to diesel. To do this, he went to his ministry, only to learn that, if he were lucky, he'd accomplish his objective in twenty years. The general manager of another amalgamation suggested to Saiken that he talk with Kabaidze: Kabaidze liked to solve people's problems. Within a few weeks, Kabaidze and his specialists had worked out a plan to enable Saiken to begin diesel engine production within two years.

Kabaidze offered his flexible machining systems as a solution. Like all Kabaidze's customers, the Zil Works received flexible manufacturing systems customized to its needs. Vladimir Chukin, head of the flexible manufacturing systems department of the Ivanovo Engineering Design Bureau, explained their philosophy. "Any standardized design configuration of FMS can be regarded as no more than a basis for developing a specific system. We design each of them in collaboration with the user, taking into account the specific conditions and requirements. We believe that universal solutions cannot be rational for any given application." When providing customers with FMSs, Ivanovo participates in their design and engineering and supplies software as well as basic metalworking equipment and transfer systems. The Zil Works have purchased sixty machining centers grouped with flexible production cells containing seven to twelve machines each.

In addition to caring about his customers, Vladimir Kabaidze cares about quality and creativity. Not only does he spend $60 million a year abroad to buy the necessary high-quality components for his beloved machine tools, but every single incoming part, be it of Soviet or Japanese origin, gets a quality control check.

Kabaidze considers himself both an artist and a professional. Every machine tool produced in his works should be the very best that can be made. To achieve this, a deep sense of professionalism is instilled into his work force. In Kabaidze's mind, machine tool construction is the very heart of industrial society. Without machine tools, a society does not have the means to build its own implements of production. Machine tools are the genetic source code of industrial society. The performance capabilities of the machine tools determine, to a significant degree, the cost and quality of the means of production.

Kabaidze is a man of complex motivation and sensibilities. He is a gentle but tenacious adversary. He leads through inspiration rather than fear and has a deep commitment to youthful energy, enthusiasm, and ways of thinking. He is an admitted social Darwinist who believes deeply that only through the severest competition and struggle is one made strong and good at what one does. Loyalty is a reciprocal commodity, and anyone who leaves Ivanovo's employment is never welcome back.

Kabaidze's respect for youth is more than a matter of a person's years. Young people can be old, and old people can be young. Youthfulness is an attitude—an attitude of innocence, enthusiasm, and not knowing what can't be done. Nostalgia for "the way it was" is a sure sign of mental aging and, at Ivanovo, unfitness for responsibility.

In addition to building strong relationships with powerful industrial customers, Kabaidze's success in battling the bureaucracy came from his "cadres." Cadres are the only asset he will never trade. Products, technology, know-how; these are for sale. But his people, especially his young people, are precious treasure into which he pours much effort and training. A close judge of character, Kabaidze looks first and foremost for "burning eyes," as he puts it. "Burning eyes" is his code for desire, right attitude. "Burning eyes" enabled Kabaidze to rewrite the script for Ivanovo's role in the machine tool production plan back in 1970 when everyone said it couldn't be done.

When Gorbachev gave Kabaidze the right to conduct foreign trade directly, in a manner similar to that used by any American company, there was concern about lack of experience. It was suggested that experienced foreign trade hands from the ministry be hired. Kabaidze demurred. He intentionally selected young, inexperienced people to develop Ivanovo's export business. He did not want to perpetuate old ways.

The average age of his design bureau, famous for its creativity in machine design, is thirty-three: "the same age as Jesus when he died," Kabaidze noted as he sipped his beer and nibbled shrimp. When he started an important new project to develop a laser-based machining center, Kabaidze selected a twenty-eight-year-old with burning eyes. When Kabaidze asked him whether he was willing to work night and day on the job to succeed, or die doing it, his young lion eagerly took on the assignment. If he had responded, "Yes, Mr. Director, I'd be honored, but will you pay me overtime or give me a raise if I succeed?" Kabaidze would have known he'd selected the wrong person. Kabaidze is quick to admit that detection of the right flame intensity is not always easy and many mistakes are made.

"The desire to do a good job is more than a matter of money. People have to want to do a good job and that comes from inspiration, example, and peer pressure. It comes from having an organizational culture that says only the best is acceptable." The ancient Greeks have a word *arete,* often translated freely as "virtue," which means excellence at doing something. *Arete* is a quality possessed with respect to an activity: the *arete* of wrestlers, shoemakers, politicians, or machine tool builders. To excel at a task requires understanding and knowledge of the job at hand. The older, opaque Socratic expression "Virtue is knowledge" becomes intelligible when understood to mean "You can be good at your job only if you take the time to learn it thoroughly." Kabaidze has machine building *arete* because of a deep professional knowledge and love of his art. He also strives mightily to have his workers possess similar *arete.* Like the ancient Greek philosopher Heraclitus, Kabaidze views conflict and struggle as the basic law of nature. All things live by the destruction of something else: life is conflict. Peace, harmony, and permanence are ideas representing death, not life.

Life is struggle, and one gets good at one's job by being competitively fit and knowledgeable. To keep up with a changing world, youthful energy and openness to new ways are essential. These are the simple principles which have guided Vladimir Kabaidze in his struggle for excellence.

Destroy a man's altruism, and you get a savage orangutan, but if you destroy his egoism, you generate a tame monkey.

ALEXANDER HERZEN

McFyodorov

"In many ways the scene resembles any modern factory. A conveyor glides silently past five work stations. Each station is staffed by an attendant in a sterile mask and smock. The workers have just three minutes to complete their tasks before the conveyor moves on; they turn out twenty finished pieces per hour." As *Time* pointed out in its July 1985 article, everything else about the assembly line was unusual: the "workers" are eye surgeons and the work pieces are humans transported on stretchers. Most extraordinary of all, however, was that this novel automaton of surgical production by methods Henry Ford would have appreciated was being performed in Moscow.

The American medical establishment would be hard put to find a person more dedicated to the application of advanced technology and organizational efficiency to health care than

Svyatoslav Fyodorov. He is also a high-profile Russian entrepreneur in good Party standing.

My own visit with Fyodorov occurred three years later while I was in Moscow for the 1988 annual meeting of the U.S.-Soviet Trade and Economic Council. As I was unknown to him, the meeting was "brokered" through George Arbatov, director of the USA and Canada Institute, who had taken an interest in the book I was writing. The institute is a prestigious policy think tank which advises the Politburo and Central Committee. Arbatov himself is a member of the Central Committee and frequent foreign policy spokesman for the Soviet government in international forums. Having been touched by the magic wand of Arbatov, I was henceforth referred to whenever I introduced myself at Fyodorov's institute as "Arbatov? Yes?"

The Moscow Institute for Eye Micro Surgery is located in the suburbs, a good half-hour drive from the center, but is known to every cabdriver as Fyodorov's institute. It is located appropriately on a street which translates roughly into "Nowheresville": Beskudnikovsky Boulevard.

A twelve-story concrete building, the institute is clinic, hospital, and R&D center. My first problem upon entering the main door was finding someone who looked like a receptionist. All I saw were twenty-five yards of coat racks and a hustle and bustle reminiscent of Grand Central Station. I stopped someone who looked as if he might know where Professor Fyodorov's office would be and was pointed toward a wing of the building.

After going back outside and finding what looked like another entrance, I entered to find only a desk, but no receptionist or evidence of anybody who looked remotely interested in who entered the building. Already twenty minutes late, I was beginning to feel desperate. Was I again in the wrong area? I clutched at a passing nurse, who indicated that Fyodorov was up on the second floor, and that I should follow her up on the elevator. The door to Fyodorov's office led into an anteroom where two secretaries were battling a gaggle of supplicants. Both were in constant contact with their commander in chief by means of walkie-talkies. I was announced to Fyodorov: Kiser of Arbatov was here. Five minutes later, I went in to meet the generalissimo.

Fyodorov is a stocky, athletic-looking man, with a ruddy complexion and a bristly hedgehog hairstyle that might once have been a flattop. He has the air of a former midwestern halfback who has successfully moved on to a second career.

If an office is a reflection of the man occupying it, then Fyodorov's office said a lot. At first, one might think one had entered the realm of a movie director. His long boxcar of an office contained a conference table the length of a bowling lane. The sides of the room were covered by television screens. Behind his desk, in addition to a large portrait of Lenin, were two twenty-four-inch Betamax screens for showing surgical procedures requiring strong stomachs: A lens transplant is not a pretty sight close-up. To the right of his main battle station was a bank of thirty smaller TV monitors for observing operations, and farther down the same wall, two more TV screens for projecting pictures of patients' eyeballs from ophthalmic machines located near the entrance to the office. On another wall was a picture of him sitting with Castro. A large cast iron eagle was flapping its wings on a window sill. Mainly this was an office of technology and frenetic activity.

My discussion with Fyodorov was "on the run," though he said this was a quiet day compared to most. During the course of our conversation, a certificate was signed for an East German who had just completed a three-day training course, numerous documents were brought in for signature, steady communications to the front two occurred via the walkie-talkie, final inspection of patients was performed, and periodically his wife entered or left, one time emerging from a back room as Fyodorov was giving me a Betamax close-up of an eyeball going under the knife.

Fyodorov thinks big. The Interbranch Research and Production Complex for Eye Microsurgery, of which the Moscow Institute is the headquarters, combines scientific research, product engineering and manufacturing, clinical diagnosis, and surgical treatment. The "Complex" will have branches in twelve cities from Khabarovsk in Eastern Siberia to Leningrad in the North and Krasnodar in the Crimea. The planned expansion, which has been approved at a cost of 120 million rubles, will allow an increase of 200,000 procedures per year from the current 100,000.

The total staff of the Complex numbers five thousand, twelve hundred of them at the Moscow Institute. "Our goal here is to combine brains, technology, and surgical skill." The brains at the institute in Moscow are represented by about forty scientists who work in fields ranging from pharmacology to pathology and histology. The technology, or reduction to practice, is performed by fifty medical engineers who develop state-of-the-art "tools" for Fyodorov's surgeons and another twenty-five programmers who are developing software for computer control of the equipment used on the "production line," as well as programs for determining the optimum length and depth of the incisions.

As a businessman who believes in his product—the best eye treatment for the most people at the lowest cost—Fyodorov wants to spread his treatment techniques around the world by franchising clinics in the use of his methods. This requires a standardization of additional production lines, a production recipe book for treatment centers that will enable an application of the techniques developed in Moscow. "Like McDonald's," I suggested after hearing of his plans to build treatment centers in Malaysia, Kuwait, Dubai, and possibly France. "Yes, McFyodorov," he shot back, not displeased by the analogy.

The technology and know-how Fyodorov is offering encompass training in specific surgical techniques as well as approaches and methods. An intensive training course takes three days and costs one thousand dollars. The institute today trains about one hundred people from all over the world annually. Over the years, Fyodorov has trained over one hundred ophthalmologists, mostly in techniques of refractive surgery. Refractive surgery includes techniques to correct nearsightedness, farsightedness, and astigmatism, controversial techniques which Fyodorov has pioneered. Known as *radial keratotomy,* the procedure for correcting nearsightedness involves cutting the cornea along lines radiating from the pupil. The effect of these incisions is to make the cornea flatter so that it bends light coming through the lens to make images focus on the retina instead of in front, the cause of nearsightedness.

Fyodorov talks with resentment about the U.S. scientific establishment, which from the beginning took either a patronizing or dismissive attitude toward his work. In his view, the American scientists showed closemindedness and exhibited all the attributes of a tightly knit establishment not interested in outsiders with new unsettling ideas. A particularly galling occasion for him was when he attended a meeting with American ophthalmologists to discuss the safety aspects of the procedure. The leader of the discussion turned first to the inexperienced Americans, who had only a couple of operations behind them, although the master who had done thousands was present in the room. In the field of refraction surgery, safety is a function of the experience of the surgeon.

In contrast to the scientific "mafia," as Fyodorov refers to the American ophthalmological community, American businessmen have been open and receptive to his innovative work. License agreements concluded or in negotiation with Bausch and Lomb and Alcon in Dallas are signs that the U.S. industrial ECG is not flat. In 1986, Fyodorov's institute earned $2 million in hard-currency from the sale of its own technology and products. The Fyodorov-Zakharov Sputnik artificial lens, made from liquid silicone, was initially dismissed by the American scientific community. Using silicone, he was told, is like putting a bomb into someone's eye. Now the Americans have come around and are also producing silicone lenses, but Fyodorov maintains that his are better and have a higher index of refraction. Bausch and Lomb seems to agree, as the firm has acquired rights to the foldable intraocular lens first developed by the Institute for Eye Microsurgery. Bausch and Lomb has also purchased rights to a novel collagen shield developed by Fyodorov's institute which promotes rapid post-operative healing of the eye.

Management Philosophy

Fyodorov is a materialist. He believes that people who work better should live better. This formula includes the doctors who are also "workers." If doctors do a good job by more efficiently providing more effective treatment to people, they should benefit. But the aim of this efficiency is to make health care self-supporting, even if the state still pays the cost of its citizens' health needs.

To implement this concept of payment for performance, in 1988 Fyodorov proposed to the Ministry the elimination of its system of providing his clients a fixed budget regardless of the quality of work and the number of patients cured. Instead, he proposed a system of earning money for the complex based on the number of patients cured. To organize "production" to achieve self-financing, a new incentive system was introduced by Fyodorov and his colleagues. The complex receives a fixed amount per procedure ranging from 167 rubles for a simple routine operation, such as a radial keratotomy, to 326 rubles for a complex procedure such as replacement of a detached retina. The average price received per patient is 250 rubles. If the hospital can do the procedure for less and increase volume as well, the savings stay within the institute, to be used for buying equipment, financing R&D, increasing wages. In 1986, Fyodorov explained that the institute saved 12 million rubles. On the basis of the staff's decision, 300,000 rubles went to a material incentive fund, 500,000 for purchase of new equipment, and the rest to various social benefit programs for workers. The system is similar to the private health care incentives of a Kaiser Permanente, but paid for by the state. Use of profits, on the other hand, is determined by the staff, or *collective* in Soviet terminology, not just top management.

To prevent the state's paying for shoddy results and cost cutting at the expense of patient care, the institute is only reimbursed for successful operations. The success of a procedure is objectively determined by visual acuity and other functional tests performed five days after the operation. Only then is payment authorized, depending on the results of the quality control examination, which would require restoration of a certain percentage of visual function by then, depending on the case. His institute is then paid quarterly according to the number and price per successful procedure. According to Fyodorov, 99 percent of his operations are successful, an improbable claim which lowers his credibility in American scientific circles.

Production work is organized into surgical groups of twenty-five people comprising doctors, nurses, and technicians under a team contract system. The performance of each group is measured by the quantity and quality of treatment. The whole team bears the economic consequences of poor performance. Thirty-two percent of the payment received for each procedure goes into a wage fund for the whole complex. Thus, an operation costing 240 rubles would produce 76 rubles for wages. Depending on the cost of an operation, a variable percentage also goes directly to the surgical team performing the operation. For a sophisticated operation, an additional 50 rubles would be received by the team; for a simple one, 11 rubles.

Under the McFyodorov system, each team is expected to perform an incredible twenty-five hundred operations per year. This is in contrast to the average of one hundred operations per year for a typical Soviet surgeon and one thousand plus per year for an American. As a result of the new incentives associated with the team contract system, the efficiency of the institute's production in early 1987 was up 73 percent over that of the preceding year, and wages had risen for everyone by 39 percent.

The team contract system is a form of self-management. The team determines its own work procedures and system of organizing itself. Overall wage relationships are set by the employees at the complex. They have adopted their notion of a "social justice" scale. The paramedicals are the lowest-paid staff; a head surgeon cannot make more than three times a paramedical. If the surgeon wants to make 600 rubles per month, he must pay 200 to the paramedicals. The general director can only make four and a half times a paramedical wage. The system is designed to recognize more sharply than the Soviet norm that leaders should have higher knowledge and skills, therefore higher wages, while preventing the differentials from becoming excessive. This results-oriented compensation system differentiates levels of skill and responsibility. Yet, scrub nurses at the complex earn 250 rubles per month, more than the average Soviet office worker.

The assembly line approach to ophthalmic surgery is a logical outgrowth of Fyodorov's view of his job. Very simply, he runs a factory whose product is patients cured or properly treated. He calls it a "happiness" factory. A well-run factory should have the highest production rate possible with the minimum of defects. To achieve this, he employs the best obtainable technology and efficient techniques of production combined with proper incentives. A highly skilled surgeon should be used very efficiently; in practice, this means only for the critical parts of an operation: "Only 10 percent of an operation needs the surgeon, but it is during that 10 percent that 90 percent of the damage is done." Therefore, the optimal use of a surgeon results from minimizing the amount of time he spends per patient. Technology and production organization are used to this end.

To get the technology, Fyodorov, like Kabaidze, must often turn to the West to equip his factory workers with the best tools. "The main difficulty is that our complex is somewhat like an island in an ocean, amidst other organizations by no means interested in the high effectiveness of their work. For example, we want to place an order with an electronic industry enterprise to quickly manufacture an instrument. We are prepared to pay good money for it. We go there and find out that another order is something nobody else wants. They say 'We can't squeeze it into our production schedule.' In addition, when Soviet equipment is delivered, it is usually inferior." Fyodorov complains about constantly having to repair Soviet-made equipment. When he visits clinics, he sees equipment costing twenty thousand rubles standing in the corner being used as coat racks.

Fyodorov has now developed his own engineering and production facility to manufacture critical instruments as well as lenses. In addition to artificial intraocular lenses, Fyodorov's institute sells diamond-coated precision surgical instruments. Currently under development is another step in intraocular lens technology, which will include an embedded solar cell to modify the optical properties of the lens required. A controllable intraocular lens of this kind will do away with the need for spectacles altogether.

Computer programs for controlling radial keratotomy operations have been written. This software is a product of Fyodorov's effort to combine quality with automation. To determine the precise degree of surgical intervention necessary for a radial keratotomy, special calculations of the cornea's condition are required. Jointly with mathematicians, the institute developed computer programs to diagnose the condition of each patient's cornea accurately and to work out an optimal surgical strategy for the length, depth, and number of incisions required.

In addition to developing high-quality "weapons" to arm his doctors, Fyodorov's other great need is procuring a sufficient number of high-quality doctor-managers: "Without good doctor-managers, the best equipped clinics will show poor results."

Like Kabaidze, Fyodorov is looking for young doctors with burning eyes to expand his domain. He tells those going out to the provinces on talent hunts, "Find a young doctor capable of producing new ideas, burning with desire to move mountains and enthusiastic enough to be ready to work in his clinic until late at night. Only then will he be able to overcome difficulties and become a real doctor-manager and coordinator." Unfortunately, such people are in short supply. Fyodorov asked who among his staff of 160 doctors were willing to spread the technology to outlying clinics in Kiev or Rostov by organizing conveyors there too. Not one came forward. Fyodorov understands why. Such projects require tremendous struggle and desire to overcome huge difficulties. "People in the USSR are not used to pushing things. A fighting spirit has to be fostered. We mainly have obedient executors of instructions."

Early Battles

Fyodorov knows something about struggle. Born into a military family in 1927, he had ambitions to be a pilot. At age sixteen, he applied for acceptance at a flying academy in Rostov. A year later, his dream was crushed under the wheels of an earthbound tram. One day when Fyodorov was running to catch a trolley car he caught his hand in the door. Dragged under the tram, Fyodorov lost half a leg. The injury also cost him his flying career.

He then chose medicine as a career and supported himself by taking up photography. When he had to choose an area of specialization, Fyodorov selected ophthalmology as a logical outgrowth of his hobby.

Upon receiving his degree from the medical institute in Rostov, Fyodorov was sent to the provinces for several years before being given a job as head of the facility at an eye institute in the town of Cheboksary. There, he read about the work of the Englishman Harold Ridley, who in 1949 performed the first experimental artificial lens implants. Excited by Ridley's work, Fyodorov decided to try to take his idea and perfect it, a thought which caused only consternation in Cheboksary. Such problems were not even being researched in Moscow, where the technology was far superior to what is available in Cheboksary. In a word, impossible. Despite the official reaction to his ideas, Fyodorov was undeterred and proceeded to manufacture lenses in his kitchen and implant them into rabbits.

A year later, in 1960, a twelve-year-old schoolgirl arrived at the institute blind in one eye as a result of congenital cataracts. Fyodorov decided to implant one of his homemade lenses, using an ordinary microscope. Luckily for Fyodorov, his first operation was a success. This patient is now a mature woman, a schoolteacher with a son named Svyatoslav in Fyodorov's honor.

When the operation was reported in the press, it unleashed an outcry of protest from the Moscow medical establishment, who could see only a physiological crime committed against the eye. The local Cheboksary officials, who were first proud that such an operation had been pioneered in their provincial institute, changed their position and withdrew support for Fyodorov's work.

Determined to pursue his revolutionary idea, Fyodorov resigned and went to Moscow to plead his case. Although he was successful in obtaining the deputy minister of health's approval to continue, the Cheboksary authorities nevertheless offered only delay and promises for the future continuation of Fyodorov's experimentation. Fed up with local timidity, Fyodorov got a job as chief of an eye clinic at a hospital in remote Arkhangel on the White Sea.

His first two operations were unsuccessful. The lenses were not yet perfected. Fyodorov succeeded in finding a skilled craftsman in Leningrad, a man with golden hands, as they are still referred to in Russia, to make his lenses. With the golden hands as his manufacturing facility, Fyodorov was able to perfect the lenses and build up a reputation as a miracle worker. In 1967, he moved to Moscow to oversee the construction of hospital number eighty-one, which would become the future Moscow Institute of Eye Microsurgery.

Today, Fyodorov's institute is considered first-class by any standards. His automated conveyor belt system uses modern production line methods. An operation is divided into separate stages or work stations. This permits using the highly qualified doctor at only the critical stage of the operation. Even though Fyodorov says the conveyor system could have been made in the USSR, because German microscopes are the best in the world, they ordered everything from the West Germans to save time. The whole production complex cost $850,000 to make and install.

The patient is strapped to an operating table that is attached to the conveyor system. A door slides open, and the table enters the operating room, where five patients can be operated on simultaneously by five specialist surgeons, each of whom performs a stage of the operation. Each person specializes—tracing the eye, aligning the laser, making the incision. One day they treat only left eyeballs, another right eyeballs.

The chief surgeon handles the most critical stage. Each stage is timed, and as the surgeon completes his part, he presses a button and the table goes on to the next doctor. Routine operations, such as lens implantations, cataract removals, and radial keratotomies, can be performed on the conveyer system. In the future, he believes 60 percent of all operations will be performed by using automated work stations. For those who are put off by the "factory" approach to operating on patients, Fyodorov's attention to the patients' comfort is reassuring. "The atmosphere in an operating wing should be comfortable, like a pleasant outing and not a cavalry charge." Before every operation, patients are counseled and have an opportunity to discuss their anxieties. Gentle pop music is piped into the operating and recovery areas.

The operation for which Fyodorov is best known, the radial keratotomy, is also his most controversial, though no more so than the lens implants of twenty years ago which are now accepted procedures. Fyodorov sees events in historical perspective. "It's like when people started flying, the first airplanes weren't too good, and it was dangerous. But now we fly all the time and don't think about it."

For Fyodorov, performing radial Ks, as they are called, is like falling off a log. His problem, according to Dr. George Waring of Emory University in Atlanta, author of a study which evaluated the safety and efficacy of the radial keratotomy procedure, is that Fyodorov's entrepreneurial enthusiasm gets ahead of his scientific rigor in presenting his results. "What is needed is an infusion of international scientific rigor to help Professor Fyodorov analyze prospectively and quantitatively the results of the enormous volume of surgery he is doing." In other words, he tends to draw conclusions which the data do not support. The problem with this critique is that pioneers break the molds of conventional wisdom by their passion and insight. They intrinsically tend to be impatient with arguments over data. Admits Waring, "He is building such an enormous medical empire in Moscow with extensions in the Soviet Union and around the world, that I think quibbling over the details of scientific reporting probably is not very important to him."

Though Fyodorov was aware of not very successful Japanese work in radial keratotomy, the specific impulse he says, came to him by chance. In 1974, a young boy was brought to Fyodorov's clinic. His glasses had been broken during roughhousing with a friend and some splinters had entered his eye, cutting his cornea. A few days later, it turned out that the boy could actually see better than before the accident. Fyodorov reasoned that if someone could help correct shortsightedness by using his fist, then surely *he* ought to be able to do something.

Now over twenty-five thousand keratotomies have been performed at his institute, thousands by Fyodorov. Like anything, the more often one does it, the better one is likely to be at it. For this reason, Moscow remains the mecca for a safe radial K. If the incisions are too long or too deep, problems can occur. Experience and equipment are the keys to safety. Fyodorov spares nothing to get the best equipment. He believes strongly that a patient is better off with an average surgeon using the best technology than an excellent surgeon with crude instruments. "The main thing is to arm doctors. Medical people are also an army, an army which does battle in peacetime. Every day we prepare for battle, the battle of the reconnaisance, diagnosis. The training is for preparation for battle in the operating theaters."

Fyodorov is now engaged in a buyout of his Moscow clinic from the Ministry of Health, to turn the facility into one that will be worker-owned in fact rather than theory. In his quest for high-quality food for patients and staff, Fyodorov's institute is reported to be investing in local Moscow agricultural cooperatives to ensure fresh meat and vegetables that will have the McFyodorov stamp of approval on them. With the possibilities opening up in the era of *perestroika*, Fyodorov's brand of entrepreneurship, combining technical prominence and business acumen, would appear to have limitless horizons.

Fyodorov is well connected, enjoying all the privileges of the power elite, but he has earned it the hard way. His personal chauffeur and dacha, two horses, and recently acquired foreign trade privileges are for him just rewards. If he earns more than others and lives better, it is because he works more. "Socialism isn't about everyone getting paid the same—that's one of our problems. It's about being paid according to your labor."

Fyodorov enjoyed no special privileges in climbing up the

ladder of success. Having overcome the physical and mental pain of losing a leg at a young age, he has little patience for self-pity and laziness. His life was a struggle, a struggle against physical disability, a struggle to implement revolutionary surgical procedures in the face of hidebound traditional thinking, and a constant struggle to get the best "weapons" to arm his doctors.

Fyodorov is autocratic, impatient and abusive with people who cross him or do not measure up to his standard. He is also a fountainhead of creativity and a managerial innovator who makes fast decisions and gets on with the program.

Kabaidze built an empire by having the courage and imagination to rewrite the script for Ivanovo. Fyodorov built his from raw technical accomplishment, drive, and creativity. Albert Kauls expanded and built upon a tradition of independent thinking developed at the Adazhi Cooperative near Riga.

Circumstances are infinite, and infinitely combined; are variable and transient; he who does not take them into consideration is . . . stark mad—he is metaphysically mad. A statesman, judging contrary to the exigencies of the moment, may ruin his country for ever.

EDMUND BURKE

The Latvian "Mormon"

Like Fyodorov, Albert Kauls sees himself as being in the business of producing happiness for people. Although his methods are different, his philosophy is similar: to live better, one must work better.

To learn more about this Baltic entrepreneur, I arranged to visit his cooperative shortly after our introductory meeting in Washington, D.C., for Kauls was the other "industrialist," albeit agroindustrialist, at the same Dartmouth conference Kabaidze had attended. Trying to interview two "heroes of socialist labor" proved to be an impossible exercise in diplomacy, so I took a raincheck from Kauls and agreed to visit him in Riga, and did two months later.

The Baltic has always held a certain fascination for me, in part because its people are ethnologically exotic. The Estonians are related linguistically to the Finns and Hungarians, the other members of the small family of Ugro-Finns. The Latvians and Lithuanians are members of the Baltic language family, a hybrid of Slavic, Germanic, and Sanskrit influences. A third Baltic language, Old Prussian, is near extinction, though spoken by a few linguistic scholars, in the manner of Old Norse or Anglo-Saxon.

The mixed linguistic influences on the Baltic languages reflect the centuries of struggle in the region and represent a protohistory of today's East-West confrontation occurring only six hundred miles farther west. The Teutonic knights brought the sword and the cross as well as lasting German influence into the region, yet they were also a harbinger of a repetitive cycle of attack and repulse by Lithuanians, Poles, Swedes, Russians, and again Germans, each seeking to dominate the strategic Baltic coastal trading towns. The Russian occupation of today began in the eighteenth century and followed upon that of every other neighbor in the area that could lay hands upon the region.

With fantasies of exotic flora and fauna dancing in my head, the flight to Riga from Moscow brought me back to reality, the reality of Western influence. At the ticket counter at Moscow's Sheremeteva Airport en route to Riga was a troup of high school baseball players from Illinois to teach Russians "how to," technology transfer which may come back to haunt America one day. The Moscow-Riga Aeroflot plane was pulsing with various forms of hard rock music. Settling into my seat, I glanced forward into the stewards' area, where I noticed a carry-on bag with the label "SPORTSWORLD, MONTANA." While I was speculating about how the steward got himself to Montana, a young man wearing a khaki shirt with a label on it sat next to me. Yes, Montana too—with the same oval burst of stars surrounding the state's name on the steward's bag. The factory that was producing these totems apparently had penetrated a deep seam of sympathetic magic which would make the Helena Chamber of Commerce proud. As I was planning to take a two-week vacation to Montana the following month, I found this McLuhanesque global village encounter vaguely unsettling. Somehow, I was not planning to go to Riga only to find Montana on my mind.

When I arrived in Riga, I was picked out of the crowd by Nadja, a young woman from Intourist, who showed me to a minivan where I was to wait while she looked after my bags. This provided me with an opportunity to chat with my driver, who turned out to be Lithuanian but spoke good Russian. As the historic Party Congress had just ended, I asked him what he thought about *perestroika*. "We already live well; we have a high standard of living because we work hard. The Russians don't want to work—we have the same land, but ours prospers and theirs doesn't."

I decided to probe intra-Baltic rivalries by asking which of the three Baltic republics lived best. My driver had no hesitation in saying Lithuania. Of the three, he thought Latvia had the lowest standard of living. Lithuania and Estonia mind their own business, go their own way. The Latvian problem is that Latvians want power; they are the most politically oriented of the three. In different ways, history gives support to this assessment. The Latvian *streltsi* or militia, were among the most trusted supporters of the Bolsheviks prior to World War I. Latvians, along with Poles, played an important role in establishing the Cheka as an instrument of revolutionary terror, most notably V. M. Latsis, second only to Dzerzhinsky, the

Pole, as a ruthless bloodletter. The recent head of the Soviet Academy of Sciences, a significant position of power, was the Latvian Mstislav Keldysh. The implication, it seemed, was that playing power games with Moscow brought with it more unproductive and inefficient ways of doing things. Politics over economics. In an unexpected way, the success story of Kauls's Adazhi Cooperative supported this interpretation of the driver's remark.

I was picked up the next morning at the Danish modern Hotel Latvia in downtown Riga and taken to Adazhi. The cooperative takes its name from the village in which it is located, about a twenty-minute drive from the city. The young man who was sent to fetch me was one of two lawyers in the cooperative's international department. He had been hired only nine months earlier because the cooperative, or Agro-Firm, as it is officially called, was granted its own foreign trade rights in 1986. Like Kabaidze and Fyodorov, Albert Kauls turned his cooperative into a model of performance, with the result that he has obtained the privilege of conducting international commerce independently of ministerial authority. Adazhi keeps for its own account all the hard-currency it earns from foreign commerce.

The main administrative building of Adazhi is an unpretentious three-story concrete shoebox. The first thing to catch my attention upon entering Kaul's office was a round oak conference table with a well in the middle. Inside the well was a vase of pink carnations in the center of a Stonehenge of cereal boxes: Apple Squares, Raisin Bran, and their relatives. The next impression, aside from one of spacious, airy orderliness, was a notion that perhaps no one actually worked there. There was not a piece of paper in sight. The chairman's desk looked as if it had just been simonized. I had a chance to peruse his bookcases while I waited. They were filled with folk art and glossy picturebooks of Latvia. This was definitely a different kind of office. Pink carnations dotted the tables, and a large fern at the side of his desk lent the office the air of an arboretum. The only evidence that this was a place of management was the bank of touch tone phones with digital readout windows next to his desk. The office of Albert Kauls puts him squarely in the category of the "clean desk" type.

My Latvian translator, supplied by Adazhi, arrived along with a man to play a videotape about the cooperative. Though my Russian is passable, I wanted to have a high rate of return for the time I was to take of Kauls and his colleagues over the next day and a half. The raw dimensions of the video story matched what I had been told in Moscow by an agricultural expert from the Institute of USA and Canada. The cooperative covers twelve thousand hectares, or about twenty-six thousand acres. Its main agricultural activities include livestock, especially dairy cows; potatoes; and feed grains. The total turnover of Adazhi in 1987 was 58 million rubles, about $100 million at the official rate of exchange. On this turnover, Adazhi produces an 11-million-ruble profit, or a 19 percent rate of return. Whatever the true economic meaning of these numbers, Adazhi clearly performs well within the rules by which they play.

In addition to their main businesses, the Agro Firm raises fox and mink for pelts, grows chicory for ersatz coffee, and has a significant production capability—almost a third of the total turnover—for making a variety of products such as furniture, construction materials, auxiliary farm implements, and even wind surfers. To spray crops, the cooperative builds simple motorized gliders which are agile and cost-effective for the acreage they are covering.

As important as the economic indicators are for the management, Adazhi places equal emphasis on building a true community culture in which the benefits to the forty-five hundred members far exceed a healthy paycheck. The cooperative has its own savings and loan program and makes available twenty-five-year loans of up to twenty thousand rubles to build homes or buy cottages from the cooperative. Scholarships are granted to students to augment their meager state stipend for attending universities or technical schools. The cooperative has its own comprehensive sports program and modern gymnasium, which would make any midwestern high school proud. Adazhi sponsors regionwide basketball tournaments, dirt bike racing, and Greco-Roman wrestling. Plans to build the most modern hospital facility in the Soviet Union, which will include outpatient care, are under way. There were good shots of Chairman Kauls's showing First Secretary Gorbachev around and receiving accolades for doing a good job.

When Kauls finally arrived at his office, we discussed the order of the day. He had arranged to give me a tour, which would start after lunch. For the first part, he had summoned one of his key deputy directors, Mr. Levusz, who ran the livestock operation that accounted for almost one-half of the cooperative's revenues. For my part, I said I'd like to spend some time with the chairman at the end of the day to learn more about his personal background and management approach. We agreed to meet at 5:00 P.M., after which he said he was available as long as necessary.

Before adjourning for lunch, I could not resist asking about all the boxes of cold cereal on the table. Kauls had collected these during his "Dartmouth" visit to the United States, which included a trip to the famous Roswell Garth farm in Iowa and a meeting with representatives of Kellogg Corporation. "Fast breakfasts," as he called cold cereals, are a good idea. They are nutritious and convenient. They are also very popular among the Soviet diplomatic corps and elite. Persuading people to change their habits would be difficult, but he seemed to think the product would sell itself once housewives realized how easy breakfast preparation would become. Kauls also likes the American institution of the truck stop "diner," another form of relief for the weary Soviet housewife. Representatives of Kellogg were expected toward the end of the month to discuss a possible joint venture. Part of Kauls's strategic thinking is to increasingly integrate his operations from planting through to final processing. Added value is the name of the game.

Mr. Levusz, whose livestock operation was the chief breadwinner for the cooperative, is a bald mountain of a man with the look of one who had done his share of farm chores over the years. I descended with him to what appeared to be an executive dining room for either Eric the Red or, possibly, Torquemada. Located next to the wine cellar, the dining room was cool and slightly dank, possibly serving double duty as a bomb shelter. As executive dining rooms go, this was definitely different. The door was a wrought iron grill. The room itself had plain, unembellished stucco walls, late Philip the II style. Occupying most of the room was a heavy oak dining table surrounded by high, straight-backed, behemoth chairs Arnold Schwarzenegger would have difficulty bench pressing. Strips of leather padding alternately gave the impression of a Viking throne and an electric chair.

Over a meal of first-class borscht and local beef, I learned from Levusz something about the economic system at Adazhi, which is known for its high wages in relation to the rest of the country. The most important feature of the system is that people are not paid according to the number of hours they work, but on the basis of a productivity index, a system which has been in place for over ten years. The actual amount paid out to individual cooperative members is determined by the smallest units, the work brigades. Brigades, in turn, are composed into collectives, which determine the wages the main operating section heads receive. It is all based on performance. Under Levusz's tenure, milk production in cows has gone from twenty-five kilograms per year to fifty-five hundred kilograms per year average, and they are aiming to achieve seven thousand kg.

The average worker at the Adazhi earns 340 rubles per month, almost 50 percent above the national average. A substantial portion of a person's monthly wage is a bonus, which runs 40 percent of base pay. It is the bonus portion of the wage which is most subject to adjustment, depending on the performance of a "collective." The brigades also determine whether certain individuals should get a little more or less, according to their effort.

Levusz makes five hundred rubles per month as deputy director of the livestock division, or "microstructure," as it is called at Adazhi. He earns only a 10 percent monthly bonus, but at year end, he has lately been getting an additional bonus of ten times his monthly wage, or another five thousand rubles. This puts his total annual pay well above that of an academician (seventy-two hundred rubles) and on a par with that of ministers. Gorbachev himself receives only eighteen thousand rubles per year officially.

Midway through lunch, we were joined by two other gentlemen, one of whom was the director of the Technology Center of the cooperative. Also a Latvian, Mr. Pakalns was in charge of keeping Adazhi aware of innovative ideas and technological developments pertinent to their various businesses. To date, his travels had been limited to the socialist countries, but Pakalns had linked up with many of the best performing agricultural partners in neighboring countries, particularly in Hungary and Czechoslovakia. Adazhi has an active cooperative program with Sluzovice in Czechoslovakia, from whom it is acquiring embryos and embryo transplant techniques for improving their dairy herd.

Pakalns has tied in with the Red Star agricultural cooperative in Hungary near Debrecen, where he learned of innovative organizational concepts which are being implemented at Adazhi. The Hungarian cooperative had been innovative by establishing a center for disseminating the latest information about new production technology to members of a voluntary association of like-minded cooperatives. Commonplace by U.S. standards, the Hungarians' idea of organizing independently of government initiative an association of cooperatives to share and disseminate information is indicative of the passivity bred by a Communist system. For this reason, Gorbachev is trying to induce Soviet citizens to begin thinking that an activity is permitted if it is not prohibited, rather than prohibited unless specifically permitted, which has been the dominant mentality.

Of particular interest to Pakalns is the subject of potatoes. In Hungary, he learned of advances in planting potatoes developed by the Dutch, but already in use in Hungary, together with the "magnet" concept which he is organizing now in Latvia. Under this concept, Adazhi acts as an organizational and scientific center which provides technical assistance to other area cooperatives interested in potato growing. Membership is completely voluntary, but the advantages to the satellite cooperatives are obvious: the use of Adazhi's expensive seed potato storage facilities, a variety of expert consultative services for optimizing yields on the basis of a study of each cooperative's local conditions, and dissemination of up-to-date technical information.

Pakalns explained the benefits for Adazhi of all these free services to the regional potato growing community. "We want to start up a major potato processing facility. This is part of our overall plan to integrate our activities towards the processing and distribution end of the chain. We will need a large supply of potatoes from qualified suppliers. What better way to assure the quality and productivity requirements than to be the consultant for the future suppliers?"

One of the technologies which will greatly increase productivity was acquired through their connections with Red Star in Hungary, a new Dutch potato cultivator which produces, in one pass of the machine, the mounded rows in which the seedlings are planted rather than taking the usual six or seven passes to "mound up" the rows. Working with the Dutch technology has also required shifting over to the use of new chemicals supplied by Bayer of West Germany. This includes a whole regimen of biodegradable chemicals for treating the seed potatoes before and after planting.

It is precisely in the area of international cooperation that

Adazhi management was clearly feeling the positive effects of *perestroika* and their status as an Agro Firm independent from various ministries. In the past, its ministry had required Adazhi to use Ciba Geigy chemicals. Before *perestroika*, the cooperative would have required multiple levels of ministerial approvals to sign off on the kind of international wheeling and dealing in which it is now engaged. For all practical purposes, Adazhi now behaves as autonomously as any Western firm.

After lunch, the tour began. It emphasized the social side of the cooperative more than the economic. We went first to the auxiliary production center, which supplies many of the products needed by the cooperative but also is a highly profitable business center, producing everything from sledge hammers and electrical systems to polyethylene bags for storing fertilizer and furniture for the schools.

From the bag production plant, we visited a new two-story high school just completed with the help of the Adazhi construction department, building materials also care of Adazhi. In contrast to the concrete functional look of the administration buildings, the school had an imaginative modern design with a completely tiled exterior as well as tiled main interior arteries. The quality of the workmanship was noticeably superior to that of typical Soviet construction. At the ground floor entrance was a large atrium, bright and airy, with abundant flowers giving a sunken garden effect as I stepped down into a tiled "well" in the center.

A picture of a Latvian World War II sniper, decorated by Americans, I was told, hung on one of walls on the perimeter of the atrium, next to the Latvian flag: a Soviet hammer and sickle with one wavy blue fringe representing the Baltic coast. Kauls takes an intense interest in education. The motive for locating the middle school on the cooperative's property and helping in its construction is to gain more influence over the curriculum and general tone of the school. There is talk of taking it over completely and running it as another Adazhi institution to minimize contamination from the outside world of sloppiness and sloth.

After the middle school came the kindergarten. I have seen a few of the better heeled kindergartens in the United States, but nothing I have seen could remotely compare with the carefully planned comprehensiveness of the Adazhi school. If there is a temple at Adazhi, it is here, where 108 priests and priestesses pay homage to their cooperative's future.

Three hundred twenty-five children between the ages of three and six have their health and developmental needs provided in a building that is a showpiece of attention to "environment." A walkway of decoratively designed red brick leads up to the entrance, through whose doors I first glimpsed an injured bird. The bird, it turned out, belonged to a luxuriant winter garden consisting of tropical plants and a variety of birds, turtles, and fish. Aside from the pleasing first impression it creates, the winter garden is also a source of botanical instruction for the children during the winter months.

Pin neat, the kindergarten shows exceptional attention to children's needs: workshop facilities: simple wooden desks seating two with a metal "bread board" nailed in at each end for hammering; two sizes of swimming pools; parquet-floored dancing and music rooms; an outside porch where the children can eat and play in the summer when it is raining; individual sleeping cubicles for their rest period; and a full battery of specialists, including a doctor, dentist, masseuse, speech therapist, and even a barber. If both parents are members of the cooperative, kindergarten is free. If only one parent is a member, the cost of attendance is fifty rubles a month, or about 25 percent of average working person's wage. Outsiders can use the kindergarten, too, but at a cost of one hundred rubles per month, few can afford to.

After the kindergarten, I was shown the sports center, passing along the way attractive-looking apartments built by the cooperative for its members, again demonstrating an attention to quality and aesthetics not typical in the USSR. The sports center is also available for use by the outside world. It represents another subject of emphasis—the building of a sound body for the equally sound mind Adazhi is trying to develop in its citizens. Kauls mentioned later that twelve youths from Adazhi had served in Afghanistan, but fortunately none was killed. He suggested excellent physical conditioning as a possible reason for their good fortune. There are several hundred Soviet soldiers missing in action or in prison, according to Kauls. He is organizing a fund to which Adazhi will be a substantial contributor in order to ransom those remaining.

I learned later in the day after a visit to the Adazhi version of Pizza Hut that Albert Kauls is an unusual mixture of a hardheaded businessman who guides his collective with a firm, driving hand and a socially sensitive caring paterfamilias. The five thousand-member extended family which Kauls is nurturing includes five hundred pensioners, who receive an additional 60 rubles per month to supplement a meager government pension of 120 rubles. The Adahzi family is a kaleidoscope of Soviet nationalities. Seventy percent Latvian and bound together by Latvian culture, cooperative membership is sought by people from all over the country. In addition to one thousand Russians, it includes families from neighboring republics of Lithuania, Estonia, and Byelorussia. There are even a few Poles, Volga Germans from Central Asia, Jews, and two North Korean families.

The demand to join the cooperative now far exceeds the openings. The reason is clear: people live better at Adahzi and become part of a system of economic and social benefits that considerably exceed that provided by typical Soviet enterprises. It also provides a healthy outdoor life. The system is based on a strong work ethic, and the family is the working unit, bound to the cooperative by an all-embracing network of social services to care for members' needs. A direct relationship is established among the quality of work, family membership, and the overall level of benefits.

Full membership in the Adazhi family comes from having

both parents employed in the cooperative. Then all social benefits are free. Not only are whole families encouraged to work at Adazhi, but the larger, the better. To encourage large families, special bonuses are given for having children. Generous maternity leave, up to three years, is given without loss of "seniority" privileges, though this can depend on the person's overall "good citizenship" standing. The strong encouragement given to members with large families is reflected in the way the pizza parlor venture was structured.

Built along the main highway to Tallin in Estonia, which passes through the Adazhi cooperative, the restaurant provides travelers a rare opportunity to stop and eat conveniently while on the road. It was built as a straight business venture and promises to be a successful one, judging by the quality of the unusual, but tasty, sausage and pickle pizza I sampled. It is a private stock company, 51 percent owned by the Adazhi cooperative itself and the remaining 49 percent by about sixty individual shareholders of the cooperative, each of whom paid five thousand rubles per share. There is no market for these shares outside the cooperative, but investors receive a 15 percent annual dividend or 750 rubles. Low-interest loans are made available to investors at 4 percent. First chance is offered to those families with four children or more.

Forgiving loans up to 50 percent, longer vacations, paid travel expenses, and educational stipends are among the benefits that increase with the length of membership in the cooperative. As in any family, there are disciplinary sanctions too. These can range from loss of monthly bonuses to more drastic measures, such as losing one's accumulated seniority or being expelled from the cooperative, a measure which requires a vote by the members.

One of the most important yardsticks used to evaluate the performance of managers at Adazhi is the person's ability to attend to subordinates' social and personal needs. If an employee is having problems at home or needs special social or health services connected with pregnancy, divorce, death, and other traumatic events, the responsible manager is expected to help the employee with those problems. This "family spirit" of concern for off-the-job needs also makes good business sense, as troubled employees invariably are less productive. The extreme to which this culture of "caring" is taken made me mention to Kauls the Mormons, who are known for their all-encompassing interest in the welfare of their members, combined with their commitment to hard work. It was an analogy he seemed to like.

Although Kauls is unquestionably the head of the family, symbolic sharing in some of the executive privileges helps build a sense of solidarity. The black Chaika limousines are at the disposal of all co-op members for special occasions such as weddings or christenings.

Finally, there is what Kauls calls the "moral" side, which is central to Adazhi's success now and in the future. One ingredient of the moral upbringing is an international spirit: a belief that everyone can learn from others and benefit from exposure to new ways of doing things. I asked what he thought he could learn from the United States. Everything. He was especially impressed by an Illinois small family farm which raised one hundred dairy cows. They had an incredible average yield of eight thousand kilograms per cow annually, produced by a husband-and-wife team. They had worked out an extremely efficient method of organization and were invited to visit Adazhi the coming winter.

Pervading Adazhi is a "work is blessed" morality, derived from centuries of domination by others. Kauls and Pakalns attribute it to a historical Latvian work ethic in which national survival was expressed through commitment to work: work to overcome, work to drown sorrow, work to stay together.

Good Performance Is Dangerous

Adazhi, however, is more than just an expression of the Latvian work ethic. Otherwise, there would be many more

Adazhis in Latvia. Adazhi is also an expression of its unusual forty-five-year-old chairman, who has guided its fortunes since 1974. Kauls is a rugged-looking, square-jawed man with a 1950s-style black, wavy pompadour.

As we settled into our Fanta cocktails on the porch of the cooperative's guest house at the end of the day, Kauls began his story with the postwar period. Conditions were terrible. Stalinism destroyed nations—not just Latvians, but Russians too. The people who suffered most were the educated, the intelligent, and the industrious. It was reverse natural selection. Stalin created a nation of robots: do what you're told; don't ask questions.

Orphaned during the war, Kauls grew up with his aunt; went to vocational school, where he learned farm mechanics; joined the Communist youth organization, Komsomol; and started work at one of Stalin's machine tractor stations, centralized agricultural machinery depots that were supposed to serve the collective farms but were mainly aggregations of equipment that did not work. At the age of twenty-five, Kauls was appointed chairman of a two-thousand-acre cooperative and admittedly knew little about agriculture. There he learned to listen to the workers, especially the older ones. He learned to benefit from the creativity of others. After two years at the cooperative, he worked in a variety of regional Party organizations, secured a job as chairman of a larger cooperative, and finally served on the Riga District Agricultural Board. From there, he went on to be the third chairman of Adazhi since 1948.

When Kauls took over in 1974 at age thirty-one, he began to make important changes. The most important were changes in the incentive system, which was to be based on both a "material" and "moral" component. The material component was to tie pay to performance, not hours. In the old days, a worker made himself intensely unpopular by being overly productive. Each working unit or collective has a Workers Council that determines bonuses and disciplinary actions for poor performers and arbitrates decisions made by the still smaller work brigades.

A related change in incentives, the "moral component," arose from Kauls's decision to focus on the family rather than the individual as the basic working unit whose loyalty to the collective is sought. Through the family, Kauls could more effectively transmit the values of hard work, emphasis on quality, intelligence, and internationalism. Much attention is focused on education. As Palkans observed, "What we don't transmit to the parents, we'll get through to their children." Adazhi is distinguished from other hardworking Latvian cooperatives by the degree to which quality and intellect are applied to work. Kauls is acutely aware of the dearth of intellect on the land. Only through combining hard work with brains can good results be obtained. Oxen work hard but rarely produce clever labor-saving ideas.

Intelligence also requires fighting the system. Defying senseless bureaucratic regulations demands courage and political strength. From the beginning of Kauls's tenure and before, Adazhi had developed its own "intelligent" culture, a culture in which its top managers thought about directives from above and did not simply follow orders. Over the years, Adazhi has benefited from a tightly knit unity between its Chairman, the cooperative membership as a whole, and the local Party organization. Kauls and others in leadership positions at Adazhi have never accepted the view that the Party is infallible. "A basic mistake of the Party was its belief that it was always right, that it was God. Under *perestroika*, the Party is beginning to accept that it may not have all the answers—that the Party must go outside itself." I was interested to know why Kauls had avoided the the-Party-is-always-right syndrome, as he had grown up in the Stalinist era. "I was always interested in doing my job as best I could and tried to think first of my workers, not my career."

Kauls has been frequently attacked for doing what was right for his collective without regard to regulations. The sports center was criticized as too big, too luxurious. It should have been built more modestly, perhaps as a geodetic bubble. The broiler chicken houses weren't built according to regulation. The sixty-five attractive apartment flats the cooperative built for its members were not paid for from the proper building fund; the day care center was much too ostentatious; the hallways were built too wide, and the children should not have separate cubicles. One of Kauls's biggest battles occurred when he drained out a swampy area near the cooperative to create a recreational lake. The sand was sold to local construction companies for two and a half rubles per cubic meter to make cement. Party bosses, including local ones, all said that the activity was illegal and had to stop. Adazhi would have to return the money it had earned to the state. Kauls responded by stopping the supply of sand to several local construction companies. Dependent on the Adazhi sand, the companies raised a large enough commotion that the offended Party personalities backed down.

Over the years that Albert Kauls has occupied the position of chairman, many envious people have sought his removal. A dramatic *causus belli* was a thirteen-hundred-acre plot Adazhi had leased in the Ukraine to grow corn, as Latvian soil was too poor for the crop. To the huge discomforture of the neighboring collectives, they found that the traditional excuses did not seem to apply to the Adazhi-grown corn, which miraculously escaped the harmful effects on yield of poor weather and insects. The carping from the neighbors was so intense that Adazhi decided to close down its corn operation.

In Latvia itself, since 1974, net profits have grown at Adazhi from 2.8 million rubles to 18 million, resulting in average monthly salaries' rising from 170 rubles to 340 rubles, and average monthly bonuses from 10 to 60 rubles. Milk production has doubled to 5.5 tons per year, still short of the American average of 6.8 tons. Through mergers with poorly performing neighbors, Adazhi has increased its land threefold.

Kauls's indisputable success has made him enemies but also provided him with protection. He does things his own way, but he gets results his defenders can point to. Since receiving his Hero of Socialist Labor award from Secretary Gorbachev, overt opposition has died down.

The Adazhi Agro Firm has received attention in part because Kauls is said by some to be a good promoter and is ambitious for national recognition. There are now seventeen other Agro Firms in the Soviet Union that have been given independent foreign trade rights on the basis of superior performance. Just as in Hungary, the agricultural side of the economy has been the driver behind the success story there, so it is possible that a similar pattern will occur in the USSR, whose leaders have carefully studied the Hungarian reform movement.

More Heroes Needed

It is a maxim for start-up companies that the founders must be willing to give up power to gain power. Gorbachev may be the last CEO of a closely held Bolshevik venture founded in October 1917. For the company to grow, the founders must share power, delegate authority, and trust others to get the job done.

Having ventured with the people's capital, the Communist party management team is facing a common corporate problem of declining return on investment, poor morale, and stagnation. Lacking outside directors, the corporation is facing the same challenge as did another rigid, closely held company almost five hundred years earlier. The eventual price of the Catholic church's inability to reform itself was a thirty-year war to regain lost territory and ultimate loss of half of Christendom. If the Party can nail the right theses to the Kremlin Wall, they may create loyal entrepreneurs working within a new corporate culture.

By singling out Kauls and Kabaidze as Heroes of Socialist Labor, Gorbachev is signaling the Party that it needs independent-minded performers, not robots and servants. That Kauls, Fyodorov, Kabaidze, and even von Ardenne during the Stalinist period not only survived but prospered is evidence that the Party has had within its ranks defenders of the unorthodox; this was acceptable as long as such people were clear winners, were loyal, and refrained from washing dirty laundry in public. The fact that Gorbachev is himself a creature of the system suggests that the Party has within itself the potential for self-renewal.

However, for entrepreneurs such as Kabaidze, Kauls, and Fyodorov to multiply themselves in sufficient quantity to make a real difference to the economic system, more than a tolerant attitude is required. The award of medals to mavericks like Kabaidze or Kauls sends important symbolic messages, but the Party line has changed in the past and can change again. Good today may be bad tomorrow.

Instead of medals, the current and future heroes need money and independence from the state. For rejuvenating entrepreneurial spirits to flourish, they must have access to financial resources that are not dependent upon political favor or bureaucratic norms. True *perestroika* will have to create financial markets or make possible a diversity of funding mechanisms to support the creative energy that is being unbottled. Money is power, and sharing power does not come naturally to the Party of Lenin and Stalin.

Gorbachev is trying to change the corporate culture by bringing in new people and reequipping it intellectually to compete in the modern world. *Glasnost* is an essential handmaiden of *perestroika*, for restructuring requires rethinking, which must not be inhibited by old nostrums and constraints. Whether Gorbachev succeeds will depend in large measure on how easily the talents of other Kabaidzes, Fyodorovs, and Kaulses are made available to the economic life of the society. That such people exist is a testimony to the extraordinary power and drive of their personalities as well as to a minimum of pragmatism in the system in which they so unnaturally fit. *Perestroika* will make a difference only if less extraordinary individuals can be "heroes" too. The society needs a critical mass of millions of little heroes for real change to occur. Like doctors, the political engineers of *perestroika* are trying to design controlled release mechanisms for the introduction of therapeutic human initiative into the body politic. But experiments with a new drug require the correct dosage to avoid the dangers of killing the patient, creating a violent reaction, or of accomplishing nothing.

In a society as historically rife with potentially disintegrative forces as the Soviet Union, a political culture based on authority and control comes naturally. *Perestroika* and *glasnost* are attitudinal code words for the delicate task of "re-engineering" the values which make up the Soviet social "product." The Party leadership will have to show extraordinary sensitivity to the Russian people's love-hate relationship with the forces of authority and order. For the country's creative energies to be unharnessed, a greater tolerance for disorder and confusion must be fostered without causing a conservative reaction from the very people who are supposed to benefit from the process of change. The boundary line between constructive change and chaos will be different for a Russian, a Georgian, a Latvian, or an American observer. It is in this dangerous borderland full of creative and destructive potential that Gorbachev is treading.

Chapter 5

Gorbachev and the T-34 Tank

We believe universal solutions cannot
be rational for any given application.

V. N. CHURKIN, IVANOVO
MACHINE TOOL WORKS

The Hungarian-born political economist Friedrich von Hayek argued in his 1944 classic *The Road to Serfdom,* that economic freedom is indispensible for political freedom. The vanguard reformers in the Soviet Union and Hungary have the challenge of finding out how much economic freedom can be allowed the people without granting true political freedom.

I take it as a given that orderly change in the Soviet Union, and by extension in Eastern Europe, requires the Party to remain the master link in the political process. History does not provide much precedent of politicians' willingly giving up power.

An important clue to whether Gorbachev and other Communist leaders can produce greater economic vitality in their societies without significantly greater political freedom is provided by Milan Simecka, a Czech Communist of the Dubček era who fell from grace after 1968. In his thought-provoking book *The Restoration of Order*, published in 1984, Simecka explains how a system survives which daily affronts common sense. Whereas Hayek's thesis starts with the need to protect the individual from the crushing power of the state by assuring domains of economic autonomy, Simecka's analysis begins with the citizenry's total dependence upon the all-embracing power of the state, the "sole dispenser of the prerequisite of existence." The surprising stability of the system he attributes to the citizens' ability to adapt. But the process of adaptation and readaptation after 1968, a process he calls the restoration of order, was successful because socialism represents a form of convenience and simplicity for people.

> Pluralist systems require a lot of attention from the individual: it is often difficult to see the wood for the trees, and political campaigning can be confusing; a surfeit of truths can debase truth in general; it is not easy to ascertain one's actual place in society; competition makes heavy demands on one's own efficiency, and so on.
>
> In order to integrate into the new society, all the citizen had to do was to come to terms with a very few basic notions: that there is only one party of government; that there is only one truth; that everything belongs to the State which is also the sole employer; that the individual's fate rests on the favour of the State; that the world is divided into friends and foes; that assent is rewarded, dissent penalized; that it is senseless to kick against the pricks; that the State does not require the entire person, just the part that projects above the surface of public life; and that if this part accepts the sole truth, then the individual may do what he or she likes in the private sphere. The State made it clear to its citizens that it did not require them to believe wholeheartedly the arguments put out by the daily propaganda. It would happily make do with passive loyalty and acceptance of the basic rules governing relations between citizen and State: between the citizen as employee and the State as employer, and between the citizen as consumer and the State as the monopoly supplier of commodities, services, culture, social welfare, education, etc.
>
> A major incentive to adapt to the new conditions was the *awareness that there was no alternative* (author's emphasis). This awareness deepened with the passage of time and the

> gradual obliteration of any reactionary, liberal or reforming hopes that had survived among the people in the fifties due partly to the conviction of that generation of adults that regimes did not last more than ten years. Many people believed that the new system would not survive, but most of the population adapted to it.

The most important observation in Simecka's commentary about adaptation and the private sphere the state allows its citizens in return for nominal loyalty is that the incentive to conform stemmed from the awareness that "there was no alternative."

The attraction of communism, or "existing socialism," as Simecka calls it, is not unlike that which military life offers for some: an ordered, top-down existence which does not invite questioning. It provides its members with ready-made formulas, fixed expectations, and predictable routines. There is the rest of the world, and there is the Army Way. If one wants to survive, do it the Army Way.

The greatest potential danger that Gorbachev runs in managing the reform process is introducing the notion that there may be alternatives to Communist-dominated rule. Conditioning people to accept the idea of choice is dangerous for a system of government that has traditionally denied people real choices.

The need to intervene in Czechoslovakia in 1968 arose not because there was any disorder or counterrevolution, but because the formal political tidiness, the outward signs of order were being disturbed. Disorderly choices were being introduced. Simecka continues:

> The historical significance of this order lies precisely in its formal existence. Of course, in everyday life, leaving the political order aside, a state of disorder, obvious to all, is the rule. There is disorder in all the various functions of social institutions, disorder in the running of the economy, in supply, services, human relations, and so on. Disorder is no secret between the people and the government, it is a daily topic of

> conversation and is even written about in the newspapers. But so long as the actual disorder does not threaten the formal political order, its existence is tolerated as something quite normal.

In 1968, evidence of formal political disorder was growing daily. Top-ranking Party leaders started speaking off the cuff, even disagreeing in public; mass meetings were being held without approved agendas; and, most disturbing of all, people were lining up to buy newspapers, as many as four or five. "In existing socialism, one of the signs of order is that citizens buy only one paper, since they know all the others will say the same," Simecka observes.

Under Gorbachev, all the 1968 signs of "disorder" which prompted the intervention in Czechoslovakia have reappeared, this time in the citadel of "existing socialism" itself. Yet Gorbachev's greatest obstacle to economic reform and unleashing of initiative and entrepreneurial spirit—passivity—may also be his greatest political asset in bringing about change. Political freedom, like much else, has been a scarce commodity throughout Russia's long history.

People do generally want better material conditions of life and an enlargement of the latitude allowed in their private sphere. As long as the PX is full and the food is decent, soldiers are not supposed to grumble. Freedom is not part of the agenda. Should *perestroika* and *glasnost* produce more quality goods and services for consumers without Western-style democracy, the likelihood is high that most Soviet citizens will continue to adapt to a system which severely limits political choice. As nearly 50 percent of the citizens of the United States routinely do not bother to exercise their right to vote in national elections and fewer still in local ones, there is little reason to think that a people unaccustomed to having any political choices will feel deprived by not having that right offered.

The T-34 Tank

Adopting foreign ideas and practices, whether Yugoslav management techniques, American-style federalism, or Hungarian economic reforms, requires distinguishing between that which is universally applicable and that which is culturally specific. The process of applying foreign concepts and experiences to their own technical culture was implemented with great success when the Soviet Union wanted to modernize its tank force in the 1930s, a story which could have important lessons for the leadership now seeking to modernize Soviet society as a whole.

The Russian T-34 is still widely regarded as the most effective medium-size tank produced in World War II. Its construction provides an instructive model for the marriage of foreign technical concepts with both local ingenuity and designers' appreciation of the realities of Soviet manpower and manufacturing capability. The T-34 also embodies many of the strengths and weaknesses in the "Russian" approach to technical problem solving: a bias toward solutions that are incremental, minimalist, crude, but effective at performing essential tasks.

"The design shows a clear appreciation of the essentials of an effective tank, adjusted to the particular characteristics of the Russian soldiers, the terrain and the manufacturing facilities available." In his book *T-34 Russian Armor,* Douglas Orgill describes the tank as a typical product of Russian society. The designers concentrated on critical features and simplicity of production. Unspoiled by exposure to cars and modern electronics, peasant tank crews did not feel they were missing sophisticated equipment and comforts. The T-34 had only a minimum of instruments. The gearbox had three forward and one reverse gear—no more than a Chevy pickup. The turret was cramped, and the crew seats did not revolve with the turret gun as in Western tanks, requiring the operators to scramble around like hamsters when reaiming. Only the commanders had radios in their tanks, and they had to double as gunners.

Crudely constructed with rough, unfinished castings, precision-machined only where absolutely necessary, and having cramped working conditions, the T-34 nevertheless shattered German illusions of tank superiority. The German Mark III was better equipped with radios, was more comfortable, and had more sophisticated engineering, yet the Russians came out ahead on essentials. Despite its faults, the T-34 achieved an optimum balance of firepower, armor protection, and mobility, making it the most significant tank of World War II. Its great potential for increasing both armor and firepower allowed this design model to remain the standard for medium tanks into the mid-1950s.

How was such a tank designed in a supposedly unsophisticated country? The art of design, like the art of politics, lies in finding the best tradeoffs among competing needs. A skilled designer instinctively knows which tradeoffs are more or less acceptable to his constituency.

The Occam's Razor of engineering design is the KISS principle, otherwise known as "keep it simple, stupid." Designer M. I. Koshkin was a master at both keeping it simple and focusing on the essential objective—to build a tank which was uncomplicated and cheap to manufacture, simple to operate, easy to repair, and effective in battle. Everything else was secondary. The very lack of sophistication in Soviet society was an advantage for the sophisticated designer. It was easier to focus on basics. Koshkin was the Vince Lombardi of tank design.

When Soviet tank designers started in the 1920s and 1950s to look at Western models, they had already understood the basic function of a tank; it is a moving platform for a weapon. If the weapon is inadequate, the tank is doomed. The weapon's effectiveness equals its ability to deliver more punch than it receives.

During 1929, a special Soviet commission was sent abroad to study foreign armor designs and to purchase sample vehicles. In the United States they bought the rights to the combination wheel track design of W. J. Christie, whose name was given to the American T-3. One of the main advantages of the Christie tank was its speed, which suited the dimensions of the Russian front. The Russians bought two models and built a copy in Kharkov, calling it the BT-1, or "fast tank," in Russian. Between 1931 and 1938, the Russians produced eight variations of the BT-1. All shared the basic Christie suspension system and sloped armor designed to minimize the impact of antitank projectiles. But, the Christie was also being Russianized and modified according to their needs and tank philosophy.

Russian designers increased firepower by adding a heavier gun. The first diesel engine for use in a tank was tried out in a BT-1 and later improved to become the basis for the V-2 used in the T-34. The BT tanks were fitted with turrets designed to adjust to larger guns than the original specifications. The wheel or tracked Christie design option was dropped for the track alone. This would simplify production and be more realistic for Russian conditions. From the modifications to Christie's tank and experience with the BT series, Koshkin and his colleagues, Alexander Morozov and Nikolai Kuchenenko, began the design of the T-34. In 1940, two prototypes had completed a two-thousand-mile test course from Kharkov to Smolensk and back to Kharkov via Kiev.

Koshkin's goal was a medium-size tank with a better gun, one simple to produce in manufacturing facilities of varying degrees of sophistication. In the final analysis, the T-34 was the product of robust common sense, rather than inspirational genius. It combined many significant improvements which added up to an unmatched medium tank in the eyes of Western tank experts.

The T-34 embodied a novel sloped armor hull which gave it superior ballistic protection. The Koshkin team was far ahead of the British and Germans in studying the optimum angle of deflection. Since World War II, Allied ballistic tests have shown that steel plate one hundred millimeters thick, sloped at an angle of sixty degrees, is equivalent to a vertical plate of three-hundred-millimeter thickness. The nose of the T-34 armor was sloped at exactly sixty degrees.

The T-34 was famous for its mobility. When German tanks became stuck in mud and snow, the T-34 more often did not. Thanks to meticulous study of the track design, the wide track of the T-34 provided a low specific ground pressure of 0.75 kilogram per square centimeter, compared to 0.95 to 1.0 kilogram per square centimeter of German and Allied medium tanks. The first use of a V-2 diesel engine gave the T-34 great power while increasing fuel reserve and lessening danger of fire. German and Allied tanks all used gasoline engines, making them self-propelled Molotov cocktails.

Koshkin never saw his tanks perform in battle. The T-34s were to go far in making German's tank force obsolete and demoralizing its troops, accustomed to fighting behind superior armor. In September 1940, Koshkin died of pneumonia just after final drawings of the T-34 were completed for mass production. The Koshkin tradition continues today as a distinctive feature of Soviet technological development, a tradition which demonstrates that effective solutions do not necessarily require the latest in technical sophistication.

Technology and Politics

Technology in the broadest sense is practical knowledge, a way of solving problems based on a particular culture and environment. Political and economic ideas migrate across international boundaries just as do scientific and technical ideas, and with the same result: transformation and adaptation to local circumstances. Where were the export controllers when they were needed to prevent dangerous Western ideas from infecting and corrupting the delicate Russian soul?

Marx was influenced by a diverse collection of Germans, Frenchmen, and Englishmen. Hegel and Feuerbach provided him with his basic intellectual tool kit: the dialectic. Saint Simon and Fourier added the scientific method and a bit of free love. The Englishman Ricardo, not Marx, was the father of the "surplus value" theory of labor exploitation. A. J. Hobson explained to Marx why capitalism had to become an international malignancy, metastasizing through colonial expansion in order to sell its ill-gotten goods. Noticeable by their absence are any important Russian contributors to Marx's *Weltanschauung*. The East-West conflict, often cast as one between irreconcilable value systems, is more accurately a West-West conflict in which the "West" prefers not to see one of its own mutated intellectual offspring.

In contrast to the world of political theory, scientists are united by the universality of scientific knowledge. Science transcends international boundaries precisely because, unlike "democracy" or "human rights," the meaning of Maxwell's equation is the same in Russia as in France. Newton's laws of motion are valid in all countries. Water is everywhere H_2O. The common language of mathematics, physics, and chemistry makes communication of results in the physical sciences easy. Basic science provides a reservoir of verifiable common knowledge to draw upon in advancing local technological needs in all countries. But unlike science, which is universal, technology is parochial. The world of technology brings us back to local history, traditions, psychology—what Edmund Burke calls "the temper and circumstances of every community," what a businessman calls "the market."

Unlike the laws of physics, the universally appealing concepts of liberty, equality, and fraternity permit no common worldwide definition but are interpreted according to local needs. And even if these needs could be defined with the rigor of a mathematical proof, every society would admix them in different proportions to suit their own temper and circumstances.

The T-34 tank represented a successful blend of Western, especially American, design concepts with Russian conditions, but the transfer of Western ideas to the Russian political environment has been more problematical. Whether fashioning a new tank or a new political system, both processes require the introduction of new values and their adaptation to local conditions. The Party leadership of the Soviet Union is faced now with the need to reengineer their society by using all the sources of social, economic, and political wisdom it can muster.

If the Party of Lenin and Stalin succeeds in the revolutionary transformation of Soviet society from a paramilitary culture into a voluntaristic one and avoids a bloodbath while maintaining a monopoly of political power, then the Politburo engineers will have made an extraordinary contribution to the world's reservoir of political know-how. In undertaking this gargantuan task, Gorbachev will have to combine the design skills of a Koskhin in applying relevant foreign political and economic models with the innovative tenaciousness and political astuteness of the successful entrepreneurs who have preceded him in creating mini-*perestroikas*.

It is often asked in the West, "How could the system produce a political phenomenon like Gorbachev?" A part of the answer lies in the West's ignorance of the pockets of originality and contrariness that have always existed in the system. The success and esteem enjoyed by innovative mavericks such as Kabaidze and Fyodorov in the Soviet Union, Wichterle in Czechoslovakia, and Birman in Hungary are explainable in terms analogous to military life.

The military analogy is also apt, for the disciplined culture of the Bolshevik party was formed by the need to survive in a deadly conspiratorial world of enemies within and without. While exiled in Switzerland, Lenin had his natural authoritarian inclinations reinforced by reading about the organizational philosophy of the Society of Jesus, articulated by St. Ignatius of Loyola. As the Pope's rapid deployment force against Protestantism, the Jesuits placed a high premium on loyalty, discipline, and strict obedience. But any good military organization values its maverick and original thinkers, if only because they are so few in number. If an officer disobeys orders, however, he had better be right. His action then may be viewed as commendable initiative. Just as a military organization cannot function if each officer can decide which orders to obey, neither will it produce first-rate officers if they are incapable of improvisation and initiative.

Military in mentality, the Party has also shown sufficient flexibility to recognize and respect productive and useful mavericks. Only now is there recognition that for the system to survive, it must change its management culture. A significant part of the change lies in recognizing that the norms of control and obedience must be mitigated to accommodate spontaneity, diversity, and more creative use of the only resource that matters: people. But only in an environment of peace and general relaxation of security consciousness will this be possible.

Reformers and Entrepreneurs

The qualities needed by entrepreneurial innovators are similar to those of political reformers, for they are fundamentally a similar breed. Both need a vision to pursue. Both have to be optimists as well as risk takers. Political reformers want to reengineer society and entrepreneurs want to engineer new products. Yet, both have the same ultimate challenge: to sell their vision and to make it work in practice. There are many good technical ideas that are never practical. Many political ideas are highly appealing but fail because they misjudge the marketplace of needs, values, and human nature.

In the postwar period, the Eastern European satellites have been in the forefront of seeking a "third way" that combines the economic security of monolithic socialism with the productive discipline and freedom of the marketplace, but doing so in the uncertain penumbra of Soviet approval. Alexander Dubček of Czechoslovakia had an intriguing vision of socialism in the mid-1960s, but the execution and timing were faulty. The lesson of the twenty-year Hungarian reform movement lies in demonstrating wider limits to Soviet tolerance for social and economic experimentation than many supposed, so long as neither the central role of the Party nor nominal loyalty to the USSR is placed in question. But it is in Communist countries that political risk takers play for higher stakes. Being politically wrong in the West means losing a job, often *in return for* a better paying one elsewhere. In the Soviet bloc, it has more typically meant execution, imprisonment, or lasting obscurity.

The entrepreneurs in this book are also experimenters as a consequence of their drive to achieve. They have behaved in unorthodox ways in their struggle against the defects and shortcomings of their own systems without rejecting them. Within their limited spheres of influence, they have made their own unconventional "third way" by creating extraordinary islands of productivity and quality long before Gorbachev made experimentation fashionable.

The Soviet Union is now declaring social and economic experimentation to be acceptable within as yet undefined limits. This will only enhance the forces for change in Eastern Europe, too, where it is most likely that "the third way" will be first found to exist or to be as illusory as the search for the Northwest Passage. The entrepreneurs whose stories have been told are the true pioneers in this quest.

Selected References

Berlin, Isaiah. 1986. *Russian Thinkers*. Harmondsworth, U.K.: Penguin Books.

Burke, Edmund. 1960. *Selected Works*. Ed. W. J. Bate. New York: Modern Library/Random House.

Cambell, Joseph. 1964. *Occidental Mythology*. New York: Viking Press.

Drucker, Peter F. 1964. *The Concept of the Corporation*. New York: Mentor Executive Library.

Feldman, Jerome. 1988. "The Hard Story of the Soft Contact Lens." *Chemtech* (September): 526–530.

Foster, Richard. 1986. *Innovation: The Attacker's Advantage*. New York: Summit Books.

Fyodorov, Svyatoslav. 1987. *Meet the Third Millennium without Spectacles*. Moscow: Novosti Press Agency Publishing House.

———. 1987. *Microsurgery of the Eye: Main Aspects*. Moscow: Mir Publishers.

Gay, Peter. 1952. *The Dilemma of Democratic Socialism*. New York: Columbia University Press.

Geneen, Harold, with Alvin Moscow. 1984. *Managing*. Garden City, N.Y.: Doubleday and Company.

Grove, Eric. 1976. *World War II Tanks*. London: Orbis Publishing, Ltd.

Halecki, Oscar. 1952. *Borderlands of Western Civilization: A History of Central Europe*. New York: Ronald Press Company.

Hayek, Friedrich A., von. 1944. *The Road to Serfdom*. London: George Routledge and Sons, Ltd.

Hegedüs, A., and G. Kosak. 1987. "Business Executives and Reform." *New Hungarian Quarterly*, no. 106 (Summer).

Herman, A. H. 1975. *History of the Czechs*. London: Allen Lane.

Johnson, Paul. 1985. *Modern Times*. New York: Harper and Row, Publishers.

Leggett, George. 1981. *The Cheka: Lenin's Political Police*. Oxford University Press.

McPhee, John. 1974. *The Curve of Binding Energy*. New York: Farrar, Straus and Giroux

Martin, Malachi. 1987. *The Jesuits*. New York: Simon & Schuster.

Naylor, Thomas. 1988. *The Gorbachev Strategy*. Lexington, Mass.: Lexington Books.

Orgill, Douglas. 1971. *T34: Russian Armor*. New York: Ballantine Books.

Pelikan, Jaroslav. 1988. "Paths to Intellectual Reunification of East and West in Europe: A Historical Typology." *IREX Occasional Papers*. Princeton, N.J.: IREX.

Peters, Thomas, and Robert Waterman. 1981. *In Pursuit of Excellence*. New York: Macmillan.

Peters, Thomas. 1987. *Thriving on Chaos*. New York: Knopf.

Sik, Ota. 1968. *Plan and Market under Socialism*. White Plains, N.Y.: International Arts and Sciences Press.

_______. 1972. *Czechoslovakia: The Bureaucratic Economy*. White Plains, N.Y.: International Arts and Sciences Press.

Simecka, Milan. 1984. *The Restoration of Order: The Normalization of Czechoslovakia*. London: Verso Editions.

Solzhenitzyn, Alexander. 1978. "A World Split Apart," Commencement Address delivered at Harvard University, June 8.

Stern, Philip. 1969. *The Oppenheimer Case*. New York: Harper and Row, Publishers.

Von Ardenne, Manfred. 1972. *Ein Glueckliches Leben fuer Technik und Forschung*. Zurich and Munich: Kindler Verlag.

Wilson, Edmund. 1983. *To the Finland Station*. New York: Farrar Straus and Giroux.

Wright, J. Patrick. 1979. *On a Clear Day You Can See General Motors*. Grosse Pointe, Mich.: Wright Enterprises.

Index

Page numbers in *italics* refer to illustrations.